Techniques, Technology and Civilisation

Marcel Mauss

TECHNIQUES, TECHNOLOGY AND CIVILISATION

Edited and introduced by

Nathan Schlanger

Durkheim Press/Berghahn Books
New York • Oxford

First published in 2006 by
Berghahn Books

www.berghahnbooks.com

Library of Congress Cataloging-in-Publication Data
A catalog record for this book is available from the Library of Congress

Mauss, Marcel, 1872-1950. [Selections. English. 2006] Techniques, technology and civilisation / Marcel Mauss ; edited and introduced by Nathan Schlanger.
p. cm.
Includes bibliographical references (p.) and index.
ISBN 1-57181-662-3
1. Social sciences--France. 2. Technology and civilization. I. Schlanger, Nathan.

H53.F7M39 2006
303.48'3--dc22

2005043523

British Library Cataloguing in Publication Data
A catalogue record for this book is available from the British Library

Printed in the United States on acid-free paper
ISBN 1-57181-662-3 hardback

Contents

Preface

The texts assembled in this volume constitute the bulk of Marcel Mauss's published statements on techniques, technology and civilisation. With the notable exception of the essay on the 'Techniques of the Body', these texts have not been previously translated into English, and nor have they been assembled and edited together. This state of affairs may in part explain their relative lack of impact across the social sciences, when compared with the other seminal writings for which Mauss is famous: on the gift, on total social facts, on primitive classification, on categories, on magical and religious phenomena, and so on.[1] In addition, the texts here assembled confirm by and large the portrayal of their author as a somewhat meandering thinker, all too prone to leave his flashes of intuition unfulfilled.[2] There are moreover some aspects of these texts that are considerably outdated, as well as uneven or partial in scope. This in turn has to do with the variable goals and circumstances of their production, which range from editorial notes and reviews for the 'in house' *Année sociologique* to conference papers, speeches to learned societies, transcripts of lecture notes, and uncorrected posthumous publications of a political nature. Lastly, it is worth keeping in mind that these texts have been conceived and written for a variety of audiences – sociologists and anthropologists, of course, but also historians (texts 7 and 11 in this volume), psychologists (texts 9 and 12), philosophers (text 11) and archaeologists (text 6). Nevertheless, granting all these caveats and cautionary notes, it remains that Marcel Mauss's writings on matters technical and technological constitute together a veritable corpus, attesting to a comprehensive, wide-ranging and genuinely pioneering achievement.

Making available this corpus in its own (albeit translated) terms should therefore represent a valuable contribution to scholarship in the social sciences. Spanning in their contents from the manufacture and exchange of stone tools in palaeolithic times, through fieldwork instructions in colonial settings, to the social and symbolic roles played by material culture (an expression which Mauss, like his contemporaries, did not use), these texts touch upon fundamental issues surrounding the sociological, anthropological and historical appreciation of techniques. The readers of this book – sociologists, anthropologists, archaeologists, technologists, students of material culture, cultural historians or historians of science – will therefore find in its pages much food for thought. Whatever the theoretical agendas and research interests they follow, they will be able with Mauss to address or reconsider such issues as material production and social reproduction, cultural traits, their dispersal and intermixing, practical skills and their social transmission, bodily practices and *habitus*, everyday life and symbolic expression, the social shaping of techniques, and so on.

Before appraising Mauss's technology in its wider scientific and ideological contexts, a few more editorial remarks on the texts here assembled are in order.

These texts – the first by Emile Durkheim*, the second by Henri Hubert*, the third co-authored with Durkheim, and the remaining nine by Mauss alone – are presented in the sequence of their writing, which is not necessarily that of their publication. Thus text 4, 'The Nation', was published posthumously from manuscript notes written by Mauss in 1919 or 1920. Likewise text 10 is the chapter on technology in the *Manuel d'ethnographie* which was published in 1947, but which is based on lectures given by Mauss in the mid 1930s.

As a rule, these texts have been translated here in their entirety, to present their author's arguments in full and avoid any 'cherry-picking' through his oeuvre. Exceptions occur when, in fairly lengthy publications, Mauss himself had delimited his relevant comments on techniques and technology under separate headings or sections, making it possible to extract them without too much loss of context or continuity (texts 4, 5 and 8). Moreover, text 5 is published in full by the Durkheim Press in *The Nature of Sociology* (Mauss 2005). Likewise, text 10 can stand on its own, since, as just noted, it is based on a lecture course dedicated to technology. The whole book from which this chapter is reproduced is also being translated and published by the Durkheim Press as *Manual of Ethnography* in 2006.

Each of these texts is preceded by a few explanatory words on the circumstances and context of its production. Brief biographical and bibliographical details will be found at the end of volume, in alphabetical order, on over twenty-five authors, mainly French, German and American, considered as main sources or protagonists for Mauss's technology. The first mention of these authors, in the Preface, in the Introduction or in Mauss's own texts, is singled out by an asterisk (*). With the exception of the previously translated 'Techniques of the Body', all the translations have been revised and key terms harmonised. As in virtually all the scientific writings of his times, Mauss made automatic and commonplace use of the masculine pronoun, and notably of the generic *homme*: translations into 'human', 'humankind' or 'humanity' have been preferred whenever possible. On the other hand, Mauss's use of the terms 'savages' and 'primitives' has been maintained, notably because these terms are often qualified, or put in quotation marks.

In the editorial texts (the Introduction, the Biographical Notes, etc.) all translations into English are my own, unless otherwise indicated.

Notes

1. For further details on Mauss and his oeuvre, see the biographical notes and bibliographical references at the end of this volume.
2. Mauss himself readily acknowledged his lack of systematic tendencies and his dislike of tedious 'writing up' (see for example Mauss 1930/1998). This was also commented on by his former students Lévi-Strauss (1950), Haudricourt (1972) and Leroi-Gourhan (1982, 2004). All too often rather narrowly considered only as Durkheim's nephew and heir, Mauss ends up being cast in some historiographical circles as an underachieving 'enigma' or 'paradox', whose life's work appears lacking in 'unity' (see Fournier 1994, Marcel 2001, Tarot 2003).

1. 'The stone age at Port Macquarie'. P.159.ACH1. New South Wales, Australia. Mounted Haddon Collection (CUMAA©). 'The search for flint deposits is typical of the entire Palaeolithic and Neolithic eras; several Australian tribes travel six hundred kilometres to fetch ochre' (Mauss, text 10).

Acknowledgements

This book has been quite a while in the making, since Bill Pickering and I first discussed it in 1998 when I was a Junior Research Fellow at St Anne's College, Oxford. Over the years, Bill has been a model of encouragement, providing guidance and tolerance in equal measure, and making sure that this project came to its end, despite all the distractions thrown in my way. My warmest thanks go to him also for marshalling and providing the resources of the British Centre for Durkheimian Studies, notably with finding the translators, Dominique Lussier and Jacqueline Redding: I am grateful to them, and also to Nick Allen for his watchful eye over many translations. Permission was kindly given by Mrs Jeffrey to use the translations of the late Professor William Jeffrey, and also by Taylor & Francis to reproduce the 1973 English translation of 'Les techniques du corps'. Thanks are due to the *Année sociologique* and to the *Revue de synthèse*, and especially to Robert Mauss, who authorised on behalf of Pierre Mauss the translation and publication of these texts. Marion Berghahn has supported this project with a useful mixture of insistent patience, and Mark Stanton has been particularly helpful when needed. Teaching on Marcel Mauss in the seminar 'Archéologie des Techniques' at the University of Paris I has helped me clarify my ideas, and I have the students and audience to thank for that. My thanks go also to all those who have read and commented on parts or the whole of this work, and finally to Judith and to Tali, Ada and Eva, who are with me all the way.

2. 'Method of forcing the bark off by the use of a lever'. P.148.ACH1. New South Wales, Australia. Mounted Haddon Collection (CUMAA©). 'The body is man's first and most natural instrument. Or more accurately, not to speak of instruments, man's first and most natural technical object, and at the same time technical means, is his body' (Mauss, text 9).

A Note on the Illustrations

To the contemporary reader, the published works of Marcel Mauss and Emile Durkheim stand out in their near total absence of images, figures or plates. True to their upbringing in the classic French philosophical tradition, they seem to have seen little point in augmenting their discourse with graphic representations to enrich, complement or schematise (and sometimes be able to dispense with) verbal descriptions - and this, even when they dealt with such observable and tangible matters as aboriginal ritual ceremonies, bodily practices or the spatial distribution of objects and traits. This absence is all the more striking and even puzzling in the case of Mauss's technology. For one, although he had been lecturing on the topic at the Ecole d'ethnographie and the Collège de France for some twenty years, it is doubtful whether he actually used any visual aids in his teaching. Even more importantly, the vast majority of the sources on which he relied for his own research were systematically and at times lavishly illustrated-be they publications by the Bureau of American Ethnology, by German scholars of the cultural-diffusionist school, and closer home, by many anthropologists and historians (e.g. Figuier, Deniker, Montandon) not to mention his colleague H. Hubert and his students A. Leroi-Gourhan* and A-G. Haudricourt*.

The black and white photographs reproduced here are therefore only images that Mauss could have seen and might have used, had he been more visually inclined. Depicting aboriginal inhabitants of New South Wales in Australia (a particularly significant culture-area for French anthropology since Durkheim's *Elementary Forms of Religious Life*), these photographs are part of the A.C. Haddon collection held at the University of Cambridge Museum of Archaeology and Anthropology. Both the precise date and the circumstances in which these photographs were taken remain to be established (their original captions all include the mention of 'Agent-General for N.S.W.'). It is nevertheless clear that they belong to a single coherent series in which several individuals figure repeatedly, and also that these individuals have been deliberately posing for the camera, following a 'mise en scène' aimed at capturing the daily life of the Aboriginal group in question. I thank the University of Cambridge Museum of Archaeology and Anthropology for permission to reproduce these photographs, and particularly Robin Boast,

Curator for World Archaeology, and Gwill Owen, Photographer, for their kind assistance.

Interspersed throughout this volume, these photographs serve to illustrate and give body to some of Mauss's main insights on techniques, and more specifically on techniques of the body: it is in this spirit that I have added to the original caption of each photograph some apposite quotations from the essay on Techniques of the Body (text 9 here) or from the chapter on Technology in the Manuel d'ethnographie (text 10).

Introduction. Technological Commitments: Marcel Mauss and the Study of Techniques in the French Social Sciences

Nathan Schlanger

Besides presenting the thrust of Marcel Mauss's works on techniques, technology and civilisation, I will dedicate particular attention here to the historical and ideological dimensions of his contribution[1]. Indeed, beyond the nowadays commonplace scholarly requirement of contextualising scientific ideas and productions, this focus reflects a more specific thesis which needs to be spelt out. Contrary to what is often left to appear, technology – the study of techniques – is not some arcane, disinterested or merely 'objective' inward-looking undertaking, but rather one that is integral to the social sciences as a whole, partaking, explicitly or not, in the key challenges and debates which permeate and motivate these disciplines. Thus, much as I will expand in the coming pages on Mauss's technology in terms of its *contents*, I will also call attention to its *constitution*, to the various conditions that fostered and influenced its coming about, the questions asked, the issues tackled, the theories enlisted, the methods used, the knowledge produced. Such an outlook will admittedly complicate to some extent, but also considerably enrich and deepen, our reading of Mauss: in addition to their substantive value for understanding technical phenomena, the texts here assembled will also prove useful as 'object lessons' for appreciating the intrinsically non-technical stakes, as it were, surrounding the making of technology.

I Approaching technology

The very idea that there might be something non-technical about technology calls for a brief terminological excursus (see also Mitcham 1994; Sigaut 1994). Definitions of techniques as such are not really at issue here; almost invariably, these definitions involve considerations of materiality, artificiality, the appropriation of nature, the production of goods and the application of knowledge, usually augmented with references to society, culture or civilisation. In due course, we will appreciate how inspiring Mauss's own conception of techniques as 'traditional efficient acts' can be, drawing attention as it does to the collective context of efficient agency, and embracing together material and non-material dimensions such as magic and aesthetics (see texts 9, 10 and 12).[2]

What requires clarification at present is rather the distinction, or distinctions, commonly made between 'techniques' and 'technology'. In its prevalent Anglo-Saxon usage, 'techniques' applies ordinarily to phenomena that are primitive, traditional, small-scale, or else skilled and tacit, while 'technology' refers to those phenomena deemed modern, complex, sophisticated, knowledge-based, objective. In both scientific and common parlance, we intuitively accept essential differences between, say, basketry techniques and ballistic technology, or between the technology of the synthesiser and the technique of its player. Similarly, mentions of 'fishing techniques' and 'fishing technology' quite naturally bring to mind contrasting historical, material and socioeconomic circumstances. Indeed, the mere labelling of some phenomena as 'technical' and others as 'technological' is bound to create a divide between them, and thus to impact, however unwittingly, on their scientific perception and study.

If this distinction is not problematic enough, the duo 'techniques'/ 'technology' participates in another equally important semantic pairing. Here techniques are the object, and technology is its logos, its discourse, its disciplinary study. In this sense, technology is to techniques what musicology is to music, what climatology is to climate, or again what criminology is to crime. This particular connotation has long prevailed in the French research tradition. The treatise by Alfred Espinas* on *Les origines de la technologie* (1897), to give one example, deals not with the earliest palaeolithic implements from caves or gravel beds, but rather with the philosophers of classical Greece who first pondered the relations between *technê*, the gods and the polis (see also text 11). Mauss himself consistently followed this distinction. His unambiguously titled article 'Les techniques et la technologie' thus begun with the claim that 'In order to talk meaningfully about techniques, it is first necessary to know what they are. Now there actually exists a science dealing with techniques ... it is the science called technology' (text 12). Conversely, he noted that, 'Among the ethnologists, therefore, technology has a great and essential role which corresponds to the fundamental nature of techniques' (text 5). A similar distinction, with comparable disciplinary and ideological thrust, appears in a well-known text entitled 'La technologie, science humaine', by the scholar who was possibly Mauss's most faithful student, André-Georges Haudricourt (1964). Likewise the book by Robert Cresswell, *Prométhée ou Pandore?*, sub-titled *Propos de technologie culturelle* (1996) deals not with the technologies or technologisation of culture, but rather with the study of techniques from a cultural (anthropological) standpoint. All that being said, there is clearly no question of enforcing any terminological exclusivity in the use of 'technology' and 'techniques'. Mauss himself left outstanding quite a few ambiguities between these terms, but we will see that such inconsistencies are not really distracting or detrimental (the same goes, it is hoped, with those I have myself left here). What is important, nevertheless, is that we acknowledge this semantic array, if only because it will serve to remind us yet again that techniques do have 'a science called technology' dedicated to their study.

This is a point well worth making. The very fact that such a terminological excursus is at all necessary can only attest to the relative disciplinary and epistemological fragility of technology in the social sciences – a discipline which, all through the half century of writings here assembled, has remained rather uncertain in both status and remit. Several factors and circumstances can be mentioned to account for this state of affairs. At a high level of generality, it is certainly possible to invoke this notorious 'occidental' reticence towards techniques, towards things manual, lowly or non-intellectual, towards that which pertains to the *homo faber* and not to the *homo sapiens* (accepting for the moment these distinctions at face value: and see Mauss's discussion below). More pragmatically, a distracting factor has been the institutional 'vagrancy' of technology during the past hundred years or so, shifting between museums with their curatorial outlook and research oriented universities. And within universities, depending on time and place, one finds technology distributed amongst archaeology and anthropology departments (in the 'four field' conception), or within prehistoric archaeology, or archaeology and history, or again archaeology and history of art (in the continental version), with more recent occurrences in cultural studies or in history of science – the whole under constant threat of being repatriated to the materials science faculty where many consider it to 'properly belong'. Objections are also manifest at a theoretical or substantive level, notably upon the invocation of the supposedly objective, solid and quantifiable subject matter of technology. It would seem upon this that techniques themselves require no interpretation – such that their study would naturally pertain to the natural sciences which need only explanation – or on the contrary, and worse still in some eyes, that they do away with interpretation altogether – and thus end up excluded from those social and human sciences who primarily calls for understanding. Be that as it may, technology does not only thread an uncertain path between engineering and philosophy (as Mitcham 1994 puts it), precariously balancing overbearing objectivity and neutral irrelevance: as a discipline, it also manifests a dearth of self-conscious reflexivity of precisely the kind widespread across most contemporary social sciences. With various exceptions, many mentioned here, the study of techniques still lacks sufficient historical and critical awareness with which to grasp those essential non-technical meanings and motivations implicated in its constitution.

In this respect, and notwithstanding the classical philosophers dear to Espinas, it was in the course of the nineteenth century that technology took its relevant form and raison d'être as a social science. From Crystal Palace to the Exposition Universelle, from the Eiffel Tower to the Suez Canal, from the steam engine to the metropolitan underground, techniques became both the instrument and the embodiment of the Age of Progress (see Mumford 1934; Winner 1977; Adas 1989). At the same time, this formidable physical presence also began to generate around it a considerable body of research and reflection. Awareness of the mechanical constructions and social reconstructions entailed in industrialisation, coupled with a desire to enhance national prestige and performance, account for the eagerness of private and public bodies to invest intellectual and institutional resources in technology – in the study of techniques. Philosophers, economists, historians, engineers, and increasingly also sociologists and anthropologists, begun to launch critical and analytical investigations of such themes as the laws of progress and the patterns of historical development, the relations between nature and artifice, the degree of emancipation or dependence brought about by mass production, and indeed the role of mechanisation and commoditisation in shaping the modern condition. Among those pursuing these lines were evolutionists like Herbert Spencer and General Pitt Rivers in Britain, and particularly German scholars such as Franz Reuleaux* on the theories of machines, E. Kapp on organ projection, and of course Karl Marx on, inter alia, the means and relations of production under capitalist conditions. In France, Republican positivist faith combined with competition against 'la science allemande' contributed to the formation or enlargement of various *sociétés savantes* and *grandes écoles* (see Zeldin 1977; Fox and Weisz 1980; Paul 1985; as well as Guillerme and Sebestik 1968, Sebestik 1984 on the rise of technological consciousness across eighteenth- and nineteenth-century Europe). Interest in technological matters also reached social thinkers and evolutionist anthropologists such as Gabriel de Mortillet and Charles Letourneau of the Ecole d'Anthropologie, as well as Alfred Espinas of the University of Bordeaux.

This briefly sketched background lends further credibility to the following proposition: what motivated Marcel Mauss in his technological contribution was his aim to advance and spread his own distinct position as part of a scientific or disciplinary debate, *and also*, at an ideological level, his recognition that techniques were of such crucial importance for contemporary societies and civilisations that their dedicated study was an urgent necessity. On both these counts, and with due acknowledgements to prevailing intellectual traditions and their posterity, it will prove useful to begin this inquiry into Mauss's technology with his famous uncle and mentor, Emile Durkheim.

II Durkheim and the containment of technology

Indeed, Mauss did more than put technology back on the agenda of the social sciences – he effectively rescued this topic from the marginal position to which Durkheim had, more or less deliberately, confined it. If so, we must clearly dedicate some critical attention to Durkheim's hitherto unexamined stance towards matters technical and technological: both because Durkheim himself underwent a radical shift from initial approbation to virtual repression, and because his attitude sheds considerable light on the broader theoretical and disciplinary standing of technology in the social sciences in general. Durkheim actually had some good reasons for being concerned with techniques, especially as they related to the modern social condition whose science he was setting out to establish. Indeed we have it from Mauss that Durkheim had 'learnt to read Marx' already during his formative stay in Germany, and also that he had introduced his young nephew to the histories of industrial progress by Louis Figuier and Becquerel (text 12). Thus, at the beginnings of his career, when he was grappling with contemporary social phenomena, Durkheim manifested a definite interest in technological and materialist perspectives.

Material density and methodological materialism

This positive disposition is evident in his first major work, on the *Division of Labour in Society* (1893). Seeking to understand the nature of the bonds uniting the individual and society, Durkheim focused on transformations in the social solidarity brought about by developments in the division of labour – from the 'mechanical solidarity' typical of small-scale societies based on individual resemblance or interchangeability, to the 'organic solidarity' prevailing in modern industrial societies, based on differentiation, specialisation and mutual dependence. Such progress in social solidarity and in the division of labour, he then argued, was caused by variations in the social milieu, and more specifically by an increase in the number of individuals and the intensity of their interactions. This increasing 'moral density' was itself paralleled by growth in 'material density', a term Durkheim coined to refer to (a) the spatial distribution of societies, which become more and more concentrated as they shift in their mode of subsistence from hunting and gathering to agriculture, (b) urban life, which enables frequent and close contacts between its settled dwellers, and (c) the number and speed of the ways of communication, which obviously encourage interactions and which become more numerous and performing as societies advance (Durkheim 1893/1902:237–44).

The sociologist stood here at a crucial juncture of his argument, in a chapter entitled 'les causes', and yet he remained surprisingly uncommitted about the relations between moral and material density. Noting that material density is a 'visible and measurable symbol' of moral density, which it can therefore replace, he left unspecified the root causes behind transformations in the social milieu: 'it is useless to specify which of these [moral or material density] has determined the other; suffice to observe that they are inseparable', or again: progress in the division of labour is due to progress in social density, 'whatever the causes of the latter' (ibid.). Such disclaimers could only reinforce the materialist tone of his argument. While he probably had no intention of researching these themes any further, Durkheim certainly drew attention to such matters of 'substrate' as the spatial, demographic and technical factors implicated in the increased densification of modern societies. In other writings, he emphasised the role of material objects in the social world, as in this suggestive passage from his 1897 book *Le suicide*:

> It is not true to say that society is made up only of individuals: it includes also material things, which play an essential role in collective life. The social fact materialises itself to the point of becoming an element of the external world. For example, a given type of architecture is a social phenomenon; it is nevertheless partly embodied in houses and constructions of all sorts which, once built, become autonomous realities, independent of individuals. The same goes for the ways of communication and transport, and for the instruments and machines employed in industry or in private life which express the state of techniques, of written language, etc. at each moment of their history. Social life, as if crystallised and fixed on material supports, thus finds itself exteriorised, and it is from the outside that it acts upon us. (Durkheim 1897a:354)

An additional focus of attention had already transpired in the 1895 *Rules of Sociological Method*. As he introduced the notion of 'social morphology', Durkheim also noted that material density usually proceeded at the same pace as moral density, and could in general serve to measure it (Durkheim 1988 [1895]:206–7). He admitted, with some circumspection, that he had been mistaken to overemphasise in the *Division of Labour* the material density as the exact expression of the moral density (1988 [1901]:207). Nevertheless, this methodological aspect was particularly crucial to Durkheim's positivist epistemology, as conveyed through the polysemous notion of *chose* or 'thing'. The repeated claim whereby 'social facts are things, and should be treated as such' was meant to underscore the reality of social facts as data, as objects of knowledge, and also (as part of Durkheim's long-standing disciplinary quarrels) to affirm their independent, irreducible and coercive character vis à vis individual consciousness. However, the follow-up claims whereby this 'only conformed to their true nature', so that 'social facts have naturally and immediately all the characteristics of the thing' (1988 [1895]:120–23) could not fail to foster objectivist expectations – and also counter-reactions, as for example from Jules Monnerot who, from his Collège de sociologie perspective, castigated 'Durkheim et son école, comme jaloux des archéologues qui, eux, ont des objets matériels, vraiment des choses, à décrire ... .' (Monnerot 1946:67). Be that as it may, at this stage of reflection, Durkheim championed a sort of 'methodological materialism', in which the selection of explanatory factors and social determinants was guided by their mode of presentation, and particularly their objective and material character.

Retraction – Sorel and Labriola

At a general level, Durkheim certainly expected his propositions to raise controversy among his peers, and he relished the opportunity to further delimit and defend his sociological domain against individualist, psychological and irrationalist objections (for example from G. Tarde or H. Bergson*). In the preface to the first edition of the *Rules*, Durkheim actually granted that in making social evolution dependent on objective, spatially determined conditions, his method could well be criticised for being materialist (1988 [1895]:73). But beyond these specifically scientific skirmishes and definitions, it soon dawned on Durkheim that, in the political and ideological sphere, his ostensibly materialist position might prove far more challenging to handle.

In the prevailing *fin de siècle* climate, the relevance of science for the understanding and betterment of human affairs was coming under debate (Prochasson 1991; Hecht 2003). Nevertheless, the actual convergence between these two aspirations – the scientific and the political – was not questioned. On the contrary, it was being embodied at that very moment in a single, newly-named actor brought to the limelight by the Dreyfus affair: the 'intellectual'. Proceeding under the premise that positive rational science could secure human progress and cure contemporary ills, various new or revamped projects were being devised and advocated to address 'la question sociale'. These were proffered as simultaneously pragmatic and principled solutions for consideration by the republican political establishment, which, for its part, was busy fending off threats from both the traditional monarchist and Catholic right, and the emerging radical left. To a large extent then, the social sciences in France came to fruition in close contact with the ideological needs of the Republic, and specifically those of its mainstream advocates and custodians. The declared ambition to deal with the facts of moral life according to the methods of positive science could serve to secure consent in the values of bourgeois democracy, and also to repel perceived revolutionary threats to the social and economic order.

Durkheim was among those who fully subscribed to this mainstream republican project, but he also encountered specific difficulties related to his seemingly ambivalent political position: '[his] case was somewhat more ambiguous for he was considered a socialist in certain right-wing circles. But in emphasising the evolutionary, pragmatic and sometimes conservative aspects of his thought, Durkheim successfully reassured leading academics and politicians' (Weisz 1983:115). It was indeed urgent for Durkheim to provide such reassurances of his Republican bona fides – namely, that his scientific sociology did not equate with revolutionary socialism. In parallel with conservative accusations, the bluntest confirmation of this urgency actually came from the left, from Georges Sorel*. Attentive to the latest scientific advances, this heterodox Marxist thinker and publicist gave place of honour to a lengthy review of Durkheim's *Rules of Sociological Method* in the inaugural issue of his radical journal *Le devenir social* (1895). Sorel criticised the sociologist for his simplistic appreciation of classes and naivety in political matters, but he also acknowledged the intellectual eminence of his adversary, and added that were Durkheim to advance towards socialism, he (Sorel) would be the first to acclaim him: 'no thinker is as well prepared as he [Durkheim] is to introduce the theories of Karl Marx into higher education, for he is the only French sociologist ... capable of grasping in their historical changes the scientific laws and material conditions of [social] becoming' (Sorel 1895:179; Stanley 1981:81ff.).

Damning praise indeed! Sorel's nonconformist opinions may have had little echo in republican circles, but this apparently plausible allegation of materialist affinities – made by one sympathetic to this cause – must have been deeply galling. Not only did it misrepresent Durkheim's convictions, it also threatened to effectively undermine his claims to offer, as a man of science, reasoned and disinterested guidance for the moral and intellectual reform of French society. It was thus imperative for Durkheim that such misunderstandings be avoided, and that his more controversial or ambiguous claims be smoothed out, rephrased and, if need be, retracted.

Besides mulling over the matter in private (see below) Durkheim seized the opportunity to publicly clarify his position through a critical analysis of Antonio Labriola's *Essais sur la conception matérialiste de l'histoire* (1897) – introduced, as it happens, by Sorel. Durkheim concurred with Marxist historical materialism that the causes of social phenomena resided in factors beyond individual consciousness, and also that the root causes of historical development do not reside in 'natural' organic circumstances, but rather in the artificial milieu and resulting modes of life created by 'socially combined' members of society. However, Durkheim vehemently disagreed with the Marxist attribution of the motor source of social evolution to the state of techniques, the conditions of labour and the instruments of production, indeed the techno-economic infrastructure:

> Just as it seems true to us that the causes of social phenomena must be sought outside individual representations, so does it seem to us false to bring them down, in the last analysis, to the state of industrial techniques and to make of the economic factor the mainspring of progress ... Not only is the Marxist hypothesis unproven, it is contrary to facts which appear established. Sociologists and historians tend increasingly to come together in their common affirmation that religion is the most primitive of all social phenomena. It is from religion that have emerged, through successive transformations, all the other manifestations of collective activities - law, morality, art, science, political forms, etc. In principle everything is religious. (Durkheim 1897b/1982:172–73)

The religious turn

Durkheim's 'discovery' of the crucial importance of the religious and ideational dimensions of social life, some time between 1895 and 1898, was thus as much ideological as theoretical in its motivations and implications. Apart from individual or biographical aspects (e.g. the reading of Robertson Smith, or the death of his rabbi father, see Lukes 1973; Lacroix 1981; Pickering 1984), this notorious shift has also to do with broader sociological factors related to the institutionalisation and consolidation of sociology, including various 'constraints' surrounding this new science and its promotion. Compelling confirmation of this comes from the recently published correspondence between Durkheim and Mauss. While only Durkheim's letters have survived, they fully document the intense links uniting the utterly serious and determined uncle, acting as father-figure and intellectual mentor to his more laid-back nephew. Writing from Bordeaux in 1897, Durkheim applied the usual remonstrance to secure Mauss's collaboration for the forthcoming *Année sociologique*, and he also entwined the affective dimension of their intellectual complicity with some revealing claims on the 'matrix' of social life:

> But mostly, what I have been afraid of – for such difficulties in dealings with publishers are in themselves easy to bear – what I have mostly feared is that, following your historical and philological scruples, you were to escape me definitively or that, at the least, you were to endorse the appropriate point of view only in order to please me and while imposing such constraints upon yourself that the value of your contribution would be diminished. However, you are one of the linchpins of the combination and thoroughly essential, not only because you are in Paris, but also because, as I foresee and hope, from the *Année sociologique* will emerge a theory which, in exact opposition to historical materialism, so gross and simplistic despite its objectivist tendency, will make of religion, and no longer of economy, the matrix of social facts. The role of the person who will specifically deal with religious facts –notwithstanding that religion should be found everywhere, or rather because of this – will therefore be considerable. But for this it must be accepted with good grace, by which I mean with the sentiment of its utility. (Durkheim to Mauss, June 1897, in Durkheim 1998:71)

> Lastly, if I ask you to sacrifice a large part of your time for the coming *Année*, this is because I genuinely believe it to be worthwhile. In addition to its documentary interest, the *Année sociologique* should mark an orientation. In fact, this sociological importance of the religious phenomena is the outcome of all I have done; and this has the advantage of summarising our whole orientation in a concrete way, more concrete than the formulae I have hitherto employed. The more I think about it, the more I believe that we should align our articles so as to secure an advance in a defined direction. This then [the sociological importance of religion] is the direction that is needed. (Durkheim to Mauss, November 1897, in Durkheim 1998:91)

These passages (the 'insider' versions of the Labriola review, as it were) make it clear that Durkheim quite consciously used religion to cast aside any lingering connotations of historical materialism – all the more so by the turn of the century, when 'religion' was no longer considered an obscurantist and reactionary threat to the free-thinking Republic, but rather a legitimate object of scientific study as a potentially constructive moral and spiritual force. When plotting with his intimate associates, Durkheim could pitch religion in a sort of 'retributive' strife which simultaneously rehabilitated the sociological discipline in ideological and political terms, and served to displace and supersede the explanatory role of the economic, infrastructural and technical dimensions of social life.

Disengagements from technology 1 – the Année sociologique

This is indeed why the subsequent near-absence of technology in the Durkheimian sociological domain cannot be seen as an inadvertent oversight (because interest lies, quite legitimately, elsewhere) so much as a deliberate attempt to keep at bay and defuse a potentially subversive or counterproductive encumbrance. In the scientifico-ideological configuration of the day, the republican morality to which Durkheim aspired had simply no use for technology, in neither disciplinary nor theoretical terms.

On the former disciplinary level, it was, paradoxically enough, Durkheim himself who introduced the rubric of *Technologie* in volume 4 (1901) of the *Année sociologique* (see text 1). Given the notoriously programmatic and expansionist ambitions of the journal, such a textual foothold could well have served to launch new research directions and herald the disciplinary recognition of technology in the social sciences. Much like 'social morphology', which also became a distinct rubric just when Durkheim had relinquished it as a sociological explanation (see Febvre 1922; Andrews 1993), so did the inclusion of *Technologie* in the *Année* – in the seventh and last section, entitled *Divers*, and alongside *Sociologie esthétique* and *Linguistique* – represent more of a disciplinary manoeuvre for stalking the grounds than a genuine theoretical opening. Gone then was the erstwhile ambiguity on the causal relations between moral and material density. What most interested Durkheim in his inaugural statement was the methodological principle whereby techniques are symptomatic of a given state of civilisation, and that they relate as witnesses or markers to the societies in which they are found. To be sure, he had some doubts regarding the strength of this correlation (e.g. Durkheim 1988 [1901]:181), and he also came to consider civilisation as a sort of 'moral milieu' or supra-social individuality which was quite independent of the state or degree of technical development, as in the 'Christian civilisation' (see text 3). Nonetheless, in this introductory text what Durkheim primarily valued about techniques was their potential to serve as measurable markers of civilisation, and not their properties as social facts or constituents.

This bias is all the more evident when compared with the position pursued, or at least hinted at, in the subsequent entries to the rubric *Technologie*. The archaeologist and museum curator Henri Hubert, who was in charge of this rubric, took a substantially more sociological stance in his own introductory statement (text 2). Hubert considered tools and implements to be social things and 'veritable institutions', with their creative and inventive dimensions, and he also stressed the collective, collaborative character of technical practices. This attitude transpired in his book reviews, which included publications in ethnography (H. Balfour*, F. Boas*), in prehistoric archaeology (G. de Mortillet, O. Montelius) and also more specifically technological works, such as W.H. Holmes* on aboriginal pottery, G. Forestier on the paleo-technology of the wheel, and Charles Frémont on the evolution of copper smelting. Hubert's critical comment on the later is well representative of his outlook:

> But he [Frémont] considers these technical issues as if they were mechanical problems, unaffected by heterogeneous data, and soluble by purely individual reasoning procedures. We do not think that things are precisely thus. ... To this institution, which is an industry, are attached collective representations, religious, magical, aesthetic etc. Among the elements of the invention, there are some that go beyond the control of individual reasons – there are certain ideas of efficiency, of traditionally transmitted forms that cannot be conceived as the work of isolated individuals. It is actually for this reason that the history of an industry should be part of its theory, because the order of inventions does not necessarily conform to the logical order of the mechanical problems it comprises. (Hubert 1903:681)

Promising as they sounded, these perspectives had nevertheless little scope for expansion in the prevailing circumstances. With the publication load and ambitions of the *Année sociologique*, the rubric *Technologie* was never allocated more than a couple of pages per volume, with brief reviews and listings but no room for in-depth studies. As for Hubert and Mauss (who had also contributed to this rubric), both were in these days being kept busy, at Durkheim's behest, with their research on primitive rituals and religions.

Disengagements from technology 2 – the Elementary Forms of Religious Life

Indeed, disciplinary containment aside, this confinement of technology was also grounded in the sociologist's increasingly idealist and dualist position. In the second edition of the *Rules*, Durkheim granted that society is made not only of individuals but also of things, but he was now quick to specify that only individuals are the active elements (1988 (1901):81). With even greater consequences, he explicitly excluded technological classifications from consideration in the groundbreaking essay on primitive classification co-authored with Mauss in 1903. Designed to link ideas together and unify knowledge, primitive classifications were therefore scientific in their essential characteristics. In this, the authors explained in a lengthy footnote,

> These are very clearly distinguished from what could be called technological classifications. It is probable that man has always classified, more or less clearly, his food resources according to the procedures he used to obtain them; for example animals who live in the water, in the skies, on the ground. But for one, the groups thus formed are not linked to each other and systematised. These are divisions, distinctions of notions, and not schemes of classification. Moreover, it is evident that these distinctions are closely

> connected to practice, of which they merely express some aspects. It is for this reason that we have not discussed them in this work, where we specifically aim at highlighting the origins of the logical procedure which lies at the basis of scientific classifications. (Durkheim and Mauss 1903:82)

This abrupt demotion of allegedly unsystematic and 'merely' practical technological classifications sounded fairly definitive, at least for the senior author. Indeed, this jettisoning was consecrated (so to speak) in his *Elementary Forms of Religious Life* (1912). With its 'elementary' emphasis, this book aimed to apprehend the fundamental aspects of social life through the study of simple or primitive civilisations, supposedly still uncluttered by the historical developments and political complexities characteristic of contemporary societies. Drawing extensively on Spencer and Gillen's Australian Aboriginal ethnography, Durkheim sought to confirm that religion was an essential and permanent aspect of humanity, indeed the veritable source, the *fons et origo*, of human civilisation.

Here at last were being fulfilled the claims advanced some fifteen years earlier, in the Labriola review and the letters to Mauss. While historical materialism no longer exerted comparable scientific or ideological pressure – both the discipline of sociology and the bourgeois Republic were by then firmly established – Durkheim seems to have retained the retributive logic of yesteryear, and also a predilection for all-encompassing explanations and prime causes. Indeed he insisted that the categories of knowledge, as well as science, moral rules, law and jurisprudence – and effectively all the main social institutions – were all born out of religion: also 'techniques and technical practices, both those which insure the functioning of moral life (law, morality, fine-arts) and those which serve material life (natural sciences, industrial techniques) derive, directly or indirectly, from religion' (Durkheim 1912:319–20). While this derivation was merely asserted, Durkheim explained in a footnote that the 'indirect' referred to industrial techniques, which can be traced to religion through the intermediary of magic – as Hubert and Mauss have supposedly shown in their 1903 'General Theory of Magic'. In another footnote in the conclusion of the book (1912:598), Durkheim again asserted that all forms of social activity have been specifically related to religion. Economy was still an exception, but given the relations between techniques and magic and the convergence between the economic and religious ideas of value and of power, this derivation was bound to be confirmed: in the end, as in the beginning; 'everything is religious'.

In addition to its 'seedbed' quality, religion, with its distinction between sacred and profane, also served Durkheim to anchor his dualistic conception of human nature. Indeed his whole thesis rested on the predicate that: 'Man is double. There are two beings in him: an individual being which has its foundations in the organism and whose circle of activities is therefore strictly limited, and a social being which represents in us the highest reality in the intellectual and moral order that we can know by observation – I mean society' (1912:23). The social, moral and cognitive implications of this metaphysical conception have been much debated in the secondary literature, but it is also the case that it has far-reaching repercussions on techniques and their appreciation. In effect, Durkheim made his *homo duplex* alternate between two worlds or phases of existence, each with their distinctive technical features and material effects. In the one phase, the individual members of society were dispersed into small autonomous groups, following their organic impulses and going about their daily business of gathering, hunting, fishing, and other such utilitarian activities which apparently do not stir much passion nor break the monotony of ordinary existence. In the other phase, in between these periods of dispersal, the group came together and convened for bouts of revitalising creativity. These intensive gatherings were generalised by Durkheim from Australian Corrobori ceremonies into a distinctive feature of humankind as a whole. As he saw it, these scenes of orchestrated ritual turmoil generated collective effervescences in which the partakers transcended themselves as isolated self-centred individuals. With their minds freed to each other and to their surroundings, they could rekindle together their ideals and sentiments of social belonging (1912:307, 313, passim).

It should come as no surprise that Durkheim had virtually nothing to say, in the *Elementary Forms* or in other publications, on the technical operations and equipment (let alone the relations of production) implicated in the world of everyday organic existence. In contrast, he focused on a particular set of objects – totems – and famously argued that they constituted the tangible expression of the sentiments and representations generated in the course of extraordinary collective rituals. These collectively achieved ideas on truth, morality, obligation, indeed on society itself, are often abstract and complex, and they therefore come to be grasped by the members of the group during their collective communions as perceptible emblems or concrete objects, be they animals, plants, decorations or flags (1912:314–15). Reliance on these elected objects and their symbolic transformation is actually due to human nature, or rather human limitation: 'We can never escape the dualities of our nature and completely liberate ourselves from physical necessities: to express to ourselves our own ideas, we need ... to fix them on material things which will symbolise them' (1912:326). These 'material intermediaries' are both necessary and arbitrary: they serve as concrete magnets on which are stamped collectively generated thoughts, and they go on serving as mnemonic devices to recall and evoke these superadded sentiments when the assembly has dissolved and *homo duplex* languishes again through routine days of ordinary existence.

Transition to Mauss: beyond things that present no danger

As we know, the divide so clearly drawn by Durkheim between subsistence and symbolism still operates today, and its various expressions remain perceptible in anthropology, archaeology and material culture studies. These may well have other antecedents than Durkheim's direct legacy, and many can be traced back to a more generalised Western mind/body dualistic conception. Still, the above discussion of Durkheim's shifting attitude on matters technical and technological can serve to render these splits and their entailed interpretive frameworks more intelligible, and thus also more amenable to criticism – with, indeed, the help of Marcel Mauss.

And precisely in order to appreciate to the full the originality and implications of Mauss's contribution, and thus fulfil the critical aims of this introductory essay, it is important to realise that Durkheim's disengagement from technology converges with and echoes a certain disengagement from modernity. Here, after all, was a scholar who had begun his life's work preoccupied with problems of social solidarity and its consolidation in the face of a persistent moral malaise, a pathology or anomie due in part to changes in contemporary conditions. We now find him dedicating what turned out to be his last major work to a metaphysical fascination with arcane hypostasising rituals, and in the process transforming far-flung 'primitives' into the privileged subjects of the French social sciences (see Vogt 1976; Kurasawa 2003). Of course, Durkheim is fully entitled to have evolved and diversified his research themes and disciplinary orientations in the course of his career. But sufficient grounds have been covered in the preceding pages to suggest that this almost inexorable shift of emphasis or retrenchment from 'modern' to 'primitive' rested on and reflected important ideological underpinnings. All the more so, as we saw, in the case of techniques, whose valuable objective qualities were undermined by suspicions of latent materialism. So much so that with Durkheim techniques ended up being, on the one hand, confined to the realm of infra-social individual subsistence activities and, on the other, deflected to the realm of supra-social markers of civilisations – and thus, in either case, relegated to the very margins of the sociological domain.

Concomitantly – so my argument goes – it is to Marcel Mauss's continuous and wholehearted engagement with modernity, with the ideological currents and political movements of his times, that we can attribute his recovery of techniques and their study as a central domain of sociological and anthropological inquiry. Durkheim had of course a social and political consciousness of his own, most forcefully and notoriously expressed in his discussion of individualism and intellectuals around the Dreyfus affair, and later on through his blatantly patriotic pamphlets during the First World War. But in the long stretches of ordinary existence spanning these exalted moments of collective fervour, Durkheim's studious and austere personality took the upper hand, leading him notably to deplore his nephew's youthful distraction in political action (see Durkheim 1998). Acknowledging these admonitions, Mauss certainly worked long and hard for the disinterested good of the *Année sociologique* and the science of sociology, but he also proved remarkably steadfast in pursuing his action as a 'citizen and comrade' and indeed in his 'written interventions in the sphere of the normative' (Mauss 1930/1998:42, and see Karady 1968; Birnbaum 1972; Gane 1992; and Fournier 1994).

The socialism which Mauss endorsed from the outset was of a rational and non-revolutionary kind, inspired by Lucien Herr and Jean Jaurès, and indeed largely approved by Durkheim. At the turn of the century Mauss paid particular attention to issues of cooperative action (including hands-on work in a cooperative bakery), professional associations and syndicalism. To these were added, following the Russian revolution, questions of cosmopolitanism and internationalism. The popular press constituted the main medium of Mauss's contributions, initially through such outlets as Sorel's *Le devenir social* but soon enough through long-term collaborations with the less radical (and better distributed) *Le populaire* and *l'Humanité*, of which he was a founding member. His political writings were certainly diverse and for the most part circumstantial, but they fully expressed his commitment to a humane socialism grounded in science (see Mauss 1997). As he later explained in a remarkable passage on 'applied sociology', the discipline he had made his own could and had to contribute to a theory of political arts, both in order to justify its own 'material existence' and to argue that sociologists could not remain in their ivory tower, leaving politics to civil servants and such 'theorising bureaucrats' (Mauss 1927:238). It was actually the moral duty of social scientists to take a less purist and disinterested attitude, for indeed:

> The public will not allow us to deal exclusively with things that are facile, amusing, curious, bizarre, passé, things that present no danger, because they concern societies which are either extinct or remote from our own. The public wants studies with conclusions relevant for the present ... let us not be weary of bringing [scientific] facts into the debate. And if our practical conclusions will turn out to be meagre and hardly topical? All the more reason for us to propagate them liberally and energetically. (Mauss 1927:240, 242)

III Humanity in crisis – between *homo faber* and the *homme total*

Having dutifully read through the histories of industrial progress recommended to him by his uncle, and having no doubt pursued Durkheim's own relevant works, Mauss was undoubtedly aware of techniques and technology from early on. Besides contributing some brief reviews to the rubric *Technologie* in the *Année sociologique* (Boas, Bogoras, Van Gennep), he apparently taught a *questionnaire de technologie* at the Ecole Pratique des Hautes Etudes in 1903. More importantly, Mauss paid considerable attention to the topic in his long essay on magic, a phenomenon which, besides religion, he specifically related to techniques: both magic and techniques had practical ends, both implicated know-how, dexterity and *tours-de-main*, and both corresponded to the definition of *actes traditionnels efficaces* (Mauss 1903, passim). Just what impact his ongoing involvements with the cooperative, syndicalist and internationalist movements had on all of this remains to be established, though we can surmise that these militant concerns increased his practical and pragmatic sensibilities. Be that as it may, it was upon the First World War that technique became for Mauss something that, as he put it, 'presented danger', and that technology became a scientific discipline whose achievements were to be 'proclaimed liberally and energetically'. In chronological, conceptual and existential terms, Mauss's unprecedented interest in techniques and their study effectively emerged from his involvement in what may be conveniently called the fieldwork of modernity – that is, the life-shattering experiences of the Great War and the intense 'intellectual organisation of political passions' that ensued (as Julien Benda put it in *La Trahison des clercs* (1927:40).

On both patriotic and personal grounds, Mauss did not hesitate to leave his pacifism behind as soon as the First World War was declared, and to volunteer for the Front. The unprecedented combination of mechanised brutality and routinised anxiety that he encountered there left permanent intellectual and existential marks on him. Besides mourning his lost friends and colleagues, Mauss discussed the sentiments of fear and panic he had to endure, and his awareness of the physical and moral force of instinct which, during extreme moments, animates – or, on the contrary, discourages and isolates – the individual (see Mauss 1924a; text 9; Fournier 1994:359ff.). As Mauss was obviously not unique in this respect, an initial appreciation of the ways by which some of his contemporaries came to terms with this experience will help us understand his subsequent 'discovery' and valorisation of techniques.

Bergsonian challenges

Together with the demographic, political and economic perturbations that followed, the Great War generated a veritable outpouring of emotional, artistic or intellectual responses. These included a substantial corpus of eschatological writings specifically concerned with the causes and consequences of this catastrophic trauma. *The Decline of the Occident* by Oswald Spengler* or *The End of the Renaissance* by Nicolai Berdiaev are some of the better known titles of this literature, whose stylistic and topical diversity is underscored by the pervasive leitmotif of doubt, despondency and disintegration. The themes of techniques and machines were favourite subjects of apocalyptic discussions: now that their efficiency as agents of death and destruction was made so evident, their inescapable presence throughout all reaches of life could be seen as a tangible objectification of the moral crisis, if not one of its original causes. With the dazzled optimism of Victorian progress long abandoned, the examples of over-industrialised America and Bolshevik Russia suggested to many that humanity, for all its impressive material achievements (if not because of them), had dismally failed to improve its present condition and future prospects. To all intents and purposes, humanity appeared bent on accelerating the process of its destruction by making it the outcome of its own self-devised machinations (see Cruickshank 1982; Johnson and Johnson 1987; Winter 1995; Prochasson and Rasmussen 1996).

A particularly eloquent discussion of techniques can be found in *Les deux sources de la morale et de la religion* published by Henri Bergson in 1932. When he finally completed his book, the ailing Collège de France philosopher was past the peak of his pre war popularity, but the influence he exerted over French intellectual, artistic and political life was still considerable and multifaceted, attracting both emulating disciples and virulent opponents (Burwick and Douglass 1992, Antlieff 1993). So far as techniques were concerned, Bergson argued in *Les deux sources* that the same instrumental ingenuity and inventiveness that has made us humans into what we are, now threaten to run out of control, to wreak havoc and spread emptiness. Industrialisation and mechanisation have encouraged artificiality and luxury, widened the gap between town and country, transformed the relations between labour and capital, and, in general, fostered a growing disparity between 'the human dimension' and the enormity of its technical creations: 'In this bloated body, the soul remains as it was, too small now to fill it, too weak to direct it'. Indeed, clamoured Bergson, we desperately need a 'supplément d'âme', a *mystique* to oversee, control and moralise the *mécanique* if the world is ever to fulfil its universal vocation and become, as he asserted at the conclusion of his essay, a 'machine à faire des dieux' (Bergson 1932:327 ff.).

As this famously enigmatic phrase intimates, Bergson's argument was not limited to the theme of disruptive *machinisme* and techniques as an uncontrollable *golem*. As a complement to this classic critique, he also promoted a markedly different conception of techniques, around the notion of *homo faber*. When it first appeared in his 1907 *Evolution créatrice*, this *homo faber* could well be understood in a relatively weak sense, as a diagnostic classificatory feature to distinguish humanity from animality: 'If, to define our species, we were to restrict ourselves to what history and prehistory show us to be the constant characteristic of humanity and of intelligence, we would not say *homo sapiens*, but *homo faber*' (Bergson 1907:140). In the 1932 *Deux sources*, however, it became clear that Bergson conceived of *homo faber* in a far stronger ontological sense, as an essential and constitutional attribute of humanity itself. As he argued, civilisation is but a superficial, acquired veneer: scratch it, and you will reveal behind it natural man – a man who is naturally moral because morality comes from nature, from biology and not from reason, from lived intuition and not from detached intellect. This natural morality has as its purpose the perseverance of the original form of social life, when man lived in 'closed societies' – self-centred and belligerent societies where the 'aberration' of democracy was unknown, and where cohesion was related to the need to exclude others. Thus, Bergson proceeded throughout *Les deux sources* to diffract and divide techniques along an ethical fault-line. He attached negative, destructive and amoral properties to the techniques associated with intelligence, rationality, civilisation and modernity, and he attributed positive, revitalising and moral qualities to techniques associated with individual organic tendencies, vital forces, mystique and volition (Bergson 1932:21ff., 54, 222ff., 249).

As we can imagine, Mauss had no sympathy for this anti-technicist and anti-intellectualist stance. At a formal level, he pursued the long-standing confrontation between Durkheim and Bergson over the epistemological merits of intuitionism and objectivism. In his 1933 panorama of French sociology, for example, Mauss singled out Bergson and his latest book as an opponent of Durkheimian sociology, noting that Bergson 'relegates the facts studied by sociologists to the domain of the "closed", of the frozen [*figé*]: he reserves to psychology, to philosophy, and even to the philosophy of the mystique, all that is "open", vital, really psychic and creator in the realm of moral and religion' (Mauss 1933a:436). Over and above these reactions, Mauss also responded to the whole of Bergson's oeuvre at a deeper, substantive level – a level that, crucially, bore upon his deeply felt experiences of the First World War.

Socialising techniques

It was upon technology, and upon the study of *actes traditionnels efficaces*, that Mauss sought to overturn this mystical *homo faber* with his own rationalist and humanist conception of *l'homme total*. To begin with, Mauss had no doubts regarding the inherently social nature of techniques. Granted, the term *homo faber* had the merit of claiming a place of honour for techniques in the history of humankind. Nevertheless, this formula could be used only on the crucial condition that 'it denote, not a "creative power" which too much resembles the "dormitive force" of opium, but a characteristic feature of communal life, and not of the individual and profound life of the spirit' (text 5). Moreover, he argued elsewhere, the Bergsonian idea of creation is actually the opposite of technicity, of creation from matter which man has not created, but adapts and transforms (text 12). Thus 'the invention of the movement or the implement, and the tradition of its use, indeed the use itself' are essentially social things: what needs to be shown, Mauss went on, is 'the degree to which all social life depends on techniques' (text 5).

Without being functionalist about it, Mauss recognised in techniques social functions, precisely because they also functioned in a literal, tangible sense, as efficient actions. To substantiate this fundamental interdependence by which techniques created and mediated social relations, Mauss invoked the notion of tradition. Instead of being some sort of organic tendency, as Bergson would have it, a form of species memory to be revealed beneath the superficial veneer of civilisation, techniques are traditional because they are taught, acquired, transmitted. Learning and doing techniques takes place in a collective context, a context which forms and informs the social constitution of its practitioners: 'any traditional practice, endowed with a form and transmitted through that form, can in some measure be regarded as symbolic. When one generation hands down to the next the technical knowledge of its manual and bodily actions, as much authority and social tradition is involved as when transmission occurs through language' (text 8).

To further argue that traditional technical actions are both symbolically and physically efficient, Mauss went to the heart of the matter: the body itself. Our most 'natural' daily actions – walking, sleeping, eating – are acts that are constructed by the collective, that form part of the social make-up of the individual, that are open to approval, recognition and evaluation. Mauss was not oblivious to the biological foundation of the human being. However, what attracted his attention to the organic body is not the vitalist and teleological thrust Bergson had invested in it, but rather, more prosaically, its physiology: the coordination of articulated motions by which it functions and by which it embodies and conveys meaning. Indeed, in so far as they vary with individuals and societies, with education, fashion and prestige, these efficient bodily acts confirm the social nature of the *habitus* (text 9) – a philosophical concept revived by Mauss, and subsequently further developed by the sociologist Pierre Bourdieu to refer to the tacit and structuring social dispositions embodied in human action.

Just as Mauss undertook to socialise the organic endowment of the living body, so he attempted to restore the reasonable nature of its thinking activities. In view of Bergson's anti-intellectual conception of *homo faber*, we can better understand the links forged by Mauss between techniques and reason – and we can also appreciate how, in the process, he overcame some of the strictures inherited from Durkheim. It is now fashionable, Mauss noted, to interrogate sociology on the origins of reason, a fashion launched some twenty five years earlier with the essay on 'Primitive Classification'. But now Mauss readily took on board these 'technological classifications' which initially (see Durkheim and Mauss 1903:82) had been so summarily cast aside as unsystematic and merely practical. The categories of thought are not only religious or institutional in origin (as Durkheim had expounded in *The Elementary Forms*), but also technical – as in the case of *number* or *space*. There was therefore much more to know about the role of weaving, basketry or the potter's wheel in the origins of geometry, arithmetic and mechanical sciences. Moreover, it was evident to Mauss that these activities are social in both their material and psychological dimensions. Their understanding necessitated therefore appropriate ethnographic and contextual sensitivity, as Mauss explained through some perceptive criticisms of the rather artificial procedures and eurocentric bias of the Swiss developmental psychologist Jean Piaget:

> The Moroccan child has a technical sense and works much earlier than the European child. On some points, he thus thinks earlier and faster and differently – manually – than the children of our own good bourgeois families ... we can see then that rigorous and extensive ethnographic observations will have to be carried out, for example throughout

> North Africa, before any conclusions of some generality can be reached. (Mauss 1933b:119)

With techniques evidently implicating both shared practices and collective representations, Mauss stressed the part of knowledge and of consciousness deployed and acquired by those engaged in technical activities. To weave a fabric, to navigate a canoe, to construct a spear, to set a trap – all are actions which suppose and at the same time generate knowledge, knowledge which is practical rather than discursive in its nature, without being for that any less social (texts 5 and 10). The technical actor 'creates and at the same time he creates himself; he creates at once his means of living, things purely human, and his thought inscribed in these things. Here is true practical reason being elaborated' (text 5). Echoing Marx's conception of praxis, this statement represents an implicit criticism of Durkheim's idealist fixation, a rejoinder to Bergson's irrationalism, and, to both thinkers, an argument regarding the co-occurrence of creativity and routine in social life. Indeed for Mauss techniques are not the expression of some individual will to power, or an instrument of mastery over nature. If they are 'a tactic for living', as Spengler melodramatically put it in *Der Mensch und die Technik* (1931), then they are a tactic for living, thinking and striving in common; they are, above all, means and mediums for the production and reproduction of social life.

Techniques redeemed

Likewise the practitioners of techniques, just as they are social beings, are also as much *homo sapiens* as they are *homo faber*; they are above all 'total men'. Mauss's 'discovery' of the *homme total* as a physical, psychological and social nexus seems to have occurred alongside that of techniques, and upon similar formative experiences. Though not expressed or necessarily conceived of in these terms, his *homme total* constituted to a large extent a counterpoint to, if not a disavowal of, Durkheim's *homo duplex*. As we have gathered, it made perfect sense now for Mauss to stress the inherent difficulties of distinguishing traditional efficient acts of techniques, of magic, or of religion (see Mauss 1924a, (text 9) and Karsenti 1997). As he warned at the onset of his lectures on the topic,

> It will sometimes be difficult to distinguish techniques from: (1) the arts and fine arts, since aesthetic activity and technical activity are on a par as regards creativity. In the plastic arts, no differentiation can be established other than the one that exists in the belief system of the artist; (2) religious efficacy. The difference lies entirely in how the native conceives of the efficacy. It is therefore necessary to estimate the respective proportions of technique and magical efficacy in the native's mind (e.g. poisoned arrows). (Text 10)

Far removed from these clearly demarcated antithetical realms into which Durkheim had cast subsistence and symbolism, this entanglement provides another reason why, as Mauss repeatedly instructed, all technical activities should be observed, recorded, photographed, sampled, collected, and understood in their totality: who does what, when, with whom, how are tools used, in what sequence, how is food prepared and consumed, how is the tie worn, how does the wearing of shoes affect walking, how is the gait that ensues subject to evaluation and approbation, indeed how techniques, objects and activities function together in a manner that is both efficient and meaningful.

This all-embracing *ethnographie intensive* undoubtedly entailed much unbridled empiricism – and indeed it took more disciplined and narrowly focused efforts to make of these insights the cornerstone of contemporary cultural technology (see Balfet 1975; Cresswell 1976; Lemonnier 1980, 1992; Schlanger 2004, 2005). At the same time, as Mauss noted a propos historical materialism (in a comment that could equally apply to Durkheim with his religious insistence), such an approach certainly helped us, in methodological terms, to

> guard against the sophistry of according primacy to one or other series of social phenomena. Neither political matters nor moral matters are in any sense dominant in any society, still less the arts applied to them. In the end all these things are no more than the concepts and categories of our social sciences, which are still in their infancy … . Politics, morals, economics are simply elements of the social art, the art of living together. Once you see this, all those contradictions between ideas and dissertations about words become pointless. Social practice, that is the only material provided for the convergent action of the moralist, the economist and the legislator. (Mauss 1924b:122)

Overall, this totalising approach had thus the merit of recasting, by principle, technical objects and practices as multifaceted documents that inform at many different and at times unexpected levels. To say that 'the tool is nothing when it is not handled' is to stress that nothing can be understood if it is not put in relation to its whole, and recognised to be changing and dynamic. To follow this trajectory of transformations involved in all technical actions and gestures is thus to gain entry into a process of ongoing construction, mediation and recombination involving material, social and symbolic elements. No longer simply markers of some developmental state, or indicators of the identity of the civilisation which produced them, as they had been for Durkheim, they are now nothing less than 'the proof of the social fact' (Mauss 1947:3). Indeed, if for Durkheim 'things' were what social facts are meant to *look like* to the sociological observer, then for Mauss things were what social facts are actually *made of* for the social actor. Whereas Durkheim saw objects acquiring their symbolic significance only at the outcome of some extraordinary rituals, Mauss on the contrary appreciated that it took perfectly ordinary practical undertakings for objects and practices to be infused with social efficiency and meaning.

This distinctive outlook can indeed serve to relate Mauss's anthropology – and specifically his technology – with some avant-garde and surrealist conceptions of the 1920s and 1930s. By drawing attention to the intricate interpenetration of the technical, the symbolic, the efficient, indeed the reasonable and the arbitrary, Mauss rejected prevailing fixations with the deliberate, the aesthetic and the representational. By giving prominence to functional, representative and ordinary aspects of human existence, Mauss sought, however implicitly, to transcend the prevailing dichotomy between low and high culture. Indeed, he could well be said to have promoted, if not actually inspired, a version of what Georges Bataille has called, in his well-known review *Documents*, a 'bas matérialisme'. Nevertheless, beyond the juxtaposition of themes and authors in *Documents* (Mauss himself published there a brief appreciation of Picasso) there remain considerable differences between the author of *Les techniques du corps* and that of *Le gros orteil*, between the Institut d'Ethnologie and the Collège de Sociologie (see Bataille 1929, 1930; Clifford 1988; Pearce 2003). Much as he willingly paid extraordinary attention to the ordinary and magnified the mundane, much as he recognised the inseparability of the symbolic and the efficient, Mauss did not for all that set out to subvert rationality or to destabilise established orders and patterns of meaning.

In fact, we have already gathered that Mauss's primary intentions were rather redemptive, almost therapeutic. Durkheim had previously valued 'primitives' for their primeval properties, representing essential and universal traits of humankind. Mauss now focused on what he himself called 'so-called' primitives also for their decentring, relativising properties. Just as they can 'teach us to think otherwise than as *homo sorbonnensis* or *oxoniensis*' (text 11), so can their technical activities, the range of objects they produce and use – and which, as Mauss demonstrated in his famous essay on the Gift, they know so well to give, to receive, and to return – help us 'civilised' occidentals to expose and repel sentiments of alienation, decadence and loss of bearing enforced on us by the enormity of historical events, indeed to transcend the chimera of disintegration with an ideal of wholeness and of plenitude. In contrast, then, to the apocalyptic responses generated by the First World War, the message articulated and expressed by Mauss through his technology is resolutely positive, as if techniques – with, and not against, the instrumental rationality they embody – were a means of recovery, almost of re-enchantment. Far from being overrun by our uncontrollable techniques, far from crying out for a mystical *supplément d'âme* to alleviate our existential angst, far from glorifying in our instinctive pre-civilised will to create power, it is in our reasoned and collective *actes traditionnels efficaces* that we will find practical help and moral solace:

> It is technique which, through the development of societies, has brought about the development of reason, sensibility and will. It is technique that makes modern humans the most perfect of animals ... It is technique that makes humans equals and that worries the gods; it is undoubtedly technique that will save humanity from the moral and material crisis in which it is struggling'. (Text 4)

IV Civilisation and nation – purity and synthesis

It may appear somewhat naïve for Mauss to place such faith in the salutary potential of techniques, but we cannot dismiss such pronouncements as the inconsequential daydreams of an unreformed worshipper of progress. To do so would keep us from appreciating that Mauss promoted the humanity of techniques to redeem the war-shaken *homme total*. It would also blind us to the fact that his technology was also engaged in another 'intellectually passionate' debate, it too exacerbated manifold by the First World War and its geopolitical aftermath. In effect, the relations between techniques and civilisation were given renewed urgency by the problem of the nation, its making and its identity, its composition and its extension. In this respect, technology was important for Mauss precisely because it enabled him to convey in theoretical and empirical terms his preferred conception of the nation, as a 'civilised' synthesis of common values and shared civic duties.

Conservative views

During the interwar years, this broadly republican model of the 'participatory' nation was coming under sustained criticism from various political, intellectual and indeed anthropological quarters. Whereas the national ideal had acquired throughout the nineteenth century primarily liberal connotations, aimed to instil sentiments of liberty, equality and allegiance among the post-revolution citizens of the state, the successive upheavals of the early Third Republic (the Boulanger crisis, the Dreyfus affair) brought about a gradual redefinition of nationalism in right-wing and conservative terms, as an emotive concept involving group identity and territorial affiliation, kinship, communes, customs and ways of life. This perspective gained increasing support among several social scientists, notably those associated with the Société d'Anthropologie de Paris. Under the leadership of anthropologist and parliamentarian Louis Marin, the ethnographic and folkloristic study of the 'traditional customs' of various *ethnies* and *civilisations* took on many of the nationalist and essentialist connotations valued by reactionary thinkers. Marin published his *Questionnaire d'ethnographie* in 1925, the very year Mauss, L. Lévy-Bruhl* and P. Rivet* established the Institut d'Ethnologie: indeed, the professional rivalry between these institutions corresponded to deep-rooted ideological and political divergences over the concept of the nation and its foundations: civic and participatory, or on the contrary territorial, genealogical and indeed racial (see Lindenberg 1990; Tombs 1991; Lebovics 1992).

These differences were notably played out in the field of technology, and specifically through one of Marin's most active protégés, Georges Montandon*. This Swiss-born doctor acquired his ill-fame during the Vichy regime, when he eagerly put his expertise in physical anthropology at the service of the infamous Commissariat aux Affaires Juives and of Nazi racist policies in general (see Schneider 1990; Birnbaum 1993). But this should not detract attention from his anthropological contributions, which were actually well known and influential during the early 1930s (see text 10, and Haudricourt, personal communication, February 1996).

In the first part of this manual, entitled *Ethnographie cyclo-culturelle*, Montandon reviewed current anthropological theories, rejected Durkheiman sociology and endorsed, with some idiosyncratic modifications, the claims and achievements of the German *Kulturkreise* school. The second part, entitled *Traité d'ergologie systématique*, contained the most comprehensive survey of ethnographic techniques then available in French. Organised along the general lines of Marin's *Questionnaire*, this mostly descriptive corpus also raised occasional theoretical points. For our purpose, the most telling is undoubtedly this:

> It will be an error, in the technological domain as in any other, to believe that the customs of primitives and savages are a threshold which we have occupied and then moved on. Their ways are often different and diverging from those followed by the occidental civilisation. The indigenous procedures of fabrication, construction ... etc., are *sui generis*. When the lineage of the constructors who had the secret [of these techniques] becomes extinct, the procedures they used are definitively lost. (Montandon 1934:215)

It was probably not the eventual disappearance of 'primitive' crafts and peoples that perturbed Montandon, who took this to be biologically inevitable. His message was rather different. To begin with – and precisely because of their alleged *sui generis* nature – technical practices and products can serve as faithful markers of the extension and limits of different ethnic groups (for example linking the Ainu of Japan with Europe, see Montandon 1937). Next, it follows that the late-coming 'occidentaloid civilisation' is not only absolutely superior in technical, social and moral terms, it also owes nothing to those 'primitives' and 'savages' who preceded it or who still survive in its colonies. Montandon's extremism notwithstanding, his work appears to have expressed a prevailing world view among conservative anthropologists of the time. The processes of social evolution, of civilisation and of nation-formation were for them primarily processes of distillation, concentration and purification – in the realms of technical activities and objects, of customs and institutions, and indeed of biology and race.

Diffusionist circles

Mauss's technological engagements were of an altogether different sort, as we can imagine, and yet it must be noted at the outset that they were broadly set within the same culture-historical diffusionist paradigm as that endorsed by his conservative protagonists. To be sure, Mauss's training and scholarship rendered him sceptical of many diffusionist claims and results, as advanced by members of the German *Kulturkreise* school, F. Graebner*, L. Frobenius* and W. Schmidt*, or indeed by the 'intransigent helliocentrists' G.E. Smith, W.J. Perry and also, to Mauss's regret, W.H. Rivers*. To summarise, Mauss's main objections were that these scholars failed to combine the study of elements and of forms, that their views on the topic of imitation were far too naïve, that they overlooked the problem of refusals of innovations, and also effectively obscured the systemic and holistic character of civilisations. He also considered that these scholars enforced an unwarranted distinction between primitive and modern, and more generally that they deprived ethnology of any sociological content by reducing it to the study of areas and layers (texts 4, 7, 12, and see Mauss 1913, 1925b, c, d).

And yet, for all that, Mauss was not averse to culture-historical diffusionism as such. Besides carefully monitoring the writings of its advocates, he also dedicated considerable constructive attention to the topic in several of his own essays, as well as in his well-attended lecture courses at the Institut d'Ethnologie and the Collège de France. While expressing his misgivings, Mauss also retained many outlooks and expressions associated with diffusionism, both in theoretical terms and at a methodological level, such as the collection of series of objects and the drawing of distribution maps. He also actively encouraged his students and colleagues to engage in such studies, as for example Haudricourt on the harness and its diffusion across Asia and Europe, Rivet on the multiple origins of Native Americans, or André Leroi-Gourhan on the Reindeer civilisation around the Arctic. Indeed, writing in the late 1930s, Leroi-Gourhan felt able to compare (with admittedly his own axe to grind) the methods of Mauss, Rivet, Granet, Schmidt and Montandon – unlike the latter two who proceeded by creating concentric circles expanding from Asia outwards, Leroi-Gourhan granted that Mauss was 'playing quite honestly, beginning with the Haïda potlatch, drawing nearer and nearer until, thirty years later, one finds oneself in Brittany, all entangled in Australian, Aïnu, Patagonian or Swiss undergrowth' (Leroi-Gourhan 2004:99).

Mauss's overall endorsement may have to do with the generalised decline of the evolutionist paradigm, but it can also be related to the importance acquired by diffusionist studies in anthropology in France and abroad. Quite simply, he recognised that it was there – in the study of traits, contacts, layers and spatial distributions – that the scientific stakes of nationalist ideologies were being played out. Far from avoiding the challenge, and precisely in order to assess and contest in their own terms the various claims advanced, Mauss accepted diffusionism as the ideological battleground for the promotion and justification of national conceptions – in much the same way as the archaeologist V. Gordon Childe (1933) acknowledged the political pertinence of Gustaf Kossina's 'pots and people' cultural paradigm. The thrust of Mauss's position and the role played by technology in its elaboration clearly transpire in the following comment, castigating those who turn to 'civilisation' in order to ground their territorial claims: 'It is almost comic to see some ill-known, ill-studied folkloric elements being invoked in front of the Peace Conference as evidence that such and such nation should extend here or there, on the grounds that we can still find there such or such shape of house or some bizarre custom' (text 4).

Mauss's awareness of the political stakes surrounding anthropological studies of civilisation could not be clearer, but his negative characterisation of these folkloric facts should not mislead us as to his intentions. His argument was that once these various traits and usages are well known and studied (by adequate ethnographic and technological means), they can serve to *defuse* rather than strengthen would-be territorial claims – of the kind stirred by the War and the ill-conceived Treaty of Versailles. These studies can confirm, as he had already argued with Durkheim (text 3), that 'this thing called civilisation' is by its nature an *international* phenomenon, and that nations are not the purified products of some essential or immemorial destiny. Indeed – Mauss argued in his essay on the nation (text 4) – it is not tradition that makes the nation but rather the nation that makes tradition, an 'invention' (vide Hobsbawm and Ranger 1983) spurred by contacts and admixtures. It is simply erroneous to consider these traits and customs as 'the products of national genius by virtue of some sort of sociological vitalism'. The belief of modern nations in *their own* civilisation, inevitably considered the first and the best, is exaggerated and illusory. The same goes for the veritable 'fetishism' they have of their own literature, arts and institutions, and indeed their techniques, which they fervently believe to have invented single-handedly.

In fact, distinctive as they may be, all these traits and customs emerge from the milieu of other societies, since all societies are somehow submerged in a *bain de civilisation*. The 1929 colloquium on La Civilisation organised by Lucien Febvre* gave Mauss an opportunity to outline some implications of this conception (text 7). Well aware that L. Marin and his colleagues had made of *civilisations* the specific objects of their nationalist studies, Mauss worked against a 'simplistic history [that was] naïvely political and unconsciously abstract and nationalist' and argued that facts and phenomena of civilisations are intrinsically international, and indeed that civilisations themselves are increasingly becoming a 'hyper-social system of societies'. This 'globalisation' was not without risks of imbalance and overdependence, and Mauss indeed recorded how upon the revolution Russian peasants proved 'unable to repair or maintain even the coarsest ploughing tools, all of which were produced abroad, [and had] as a result to revert progressively to the most primitive form of agriculture' (text 4). Overall, however, his aim remained to demonstrate in concrete ways that nations and civilisations were increasingly tributary to ever-expanding transfers and interactions. As Mauss readily granted to the hyperdiffusionists Smith and Perry – but certainly with more profound implications – 'we have no systematic opposition to such studies of mixtures of civilisations, of mixtures of institutions, of this chemistry which makes of their social substrate a new society, with its new language, its new institutions, its different equipment' (Mauss 1925d:342).

This conviction is further conveyed in Mauss's admiring exclamation, a propos the diversity of styles at the temples of Angkor: '*Déjà un métissage, aussi magnifique que singulier*'. Both the borrowing of the racial term 'métissage' and its evident valorisation could not be more significant. They confirm that Mauss foresaw a free-flowing and generative mélange of an increasingly voluminous and important stock of traits and forms. This interactive *fonds commun*, now resolutely cross-cultural and all-embracing, formed a veritable *nouvelle civilisation* in which hitherto marginal or severed populations have the scope to contribute, to give as well as to receive: 'the success of the primitive arts, including music', Mauss predicted, fully attentive to the cultural movements of his times, 'demonstrates that the history of all this will follow many unknown paths' (text 7).

Unpredictable as the roads of this cultural, artistic and indeed artefactual *métissage* may be, circulation is essential, and must be maintained. What is anathema to the conservative anthropologists, who seek distinctive traits and techniques with which to establish their version of 'True France' as the inalterable essence of the *terroir*, the destiny and the blood, is for Mauss a constituent condition of the moral and material progress of the Republic – a France in which there is as much room for Josephine Baker as there is for Joan of Arc. To be sure, the internationalist conception which Mauss promoted throughout these writings fully recognised the importance of the material and ideal attachment to the nation and its symbols. Gone were the pre war days of Bolshevik-inspired 'cosmopolitanism', in which the human actor, deemed everywhere identical, transcended the realities of social life, and where class membership was supposed to brush aside national allegiances with uncompromising pacifism (see Mauss 1920, text 4). Nevertheless, sober and accommodating as it may have become, Mauss's understanding of the nation was still far removed from the 'closed' and exclusive society yearned for in conservative circles. Much as he remained suspicious of nationalisms, he maintained his trust in the practical and moral value of the nation precisely because it constituted an avenue towards an 'open', inclusive society, one that invited opportunities for dialogue and cross-fertilisation, one that valued possibilities of contacts and of exchanges.

Conclusions – humanity and humility

Techniques and their study were in this respect both a model and a confirmation of this ideal. In contradistinction to conservative conceptions, technical products and practices are not the markers of *sui generis* identity or essence, nor indeed are they objective milestones in a scale of progress; rather, 'each art must be studied in itself, without considering whether or not it is primitive' (text 10). Likewise, technical products and practices must be understood as the outcome of multiple convergences and interactions: seen in this light, diffusionist studies need not be confined to charting the victorious march of national genius from their original cradles. They can also be used to show that material and moral changes in societies across time and space are best understood in relation to the interactive milieu of other societies, in contact with each other and with their respective surroundings. Further studies can show that societies, like techniques and like the *homme total* who practises them, are made out of synthesis rather than distillation, that their lack of 'purity' is their source of strength, that *métissage* or *créolisation* is not their worst nightmare but their salutary fate. After all, 'By their nature, techniques tend to be generalised and to multiply everywhere throughout humanity. They are the most important factor in the causes, means, and the ends of what is called civilisation, and also of progress, not only social but also human' (text 5).

As we recall from his frank discussion of 'applied or political sociology' during the renewed launching of the *Année sociologique* (1927), Mauss considered it his duty as a member of society and as an intellectual to engage with modernity. Moreover, he also saw it essential to convey this engagement to all members of his contemporary society, and to explain it throughout all its groups and movements. Mauss was well aware that the conclusions of his science could appear at times arcane and immaterial. He therefore knew that his message stood its best chances of getting across as a tangible, readily perceptible demonstration. The one forum where his inspired vision could most visibly come to the fore was of course at the Musée de l'Homme – an institution revitalised in the 1930s through the cultural politics of the socialist *Front populaire*, directed by the militant anti-fascist Paul Rivet, and subsequently the base of the first *résistance* network in Nazi-occupied France. Displaying humanity's technical achievements in its newly built corridors – including series of objects, distribution maps, evidence of borrowing, interactions and the like – provided for the visiting masses a concrete moral lesson to absorb and, hopefully, to live by.

And in between visits to the museum, like-minded messages were also available to the masses in the popular press – to which, we recall, Mauss had been a committed contributor for most of his life. 'Where are the times ...', so begun his 1921 review of Marcellin Boule's influential manual *Les hommes fossiles* in the leftist daily *Le populaire*,

> Where are the times when Marx and Engels acclaimed Carl Vogt's Lessons on man, and when they were fascinated by Lewis Morgan's discoveries, keeping abreast of the latest publications on the Origins of the Family? Where are the times when socialism was scientific not only by name; but in seeking to integrate science and politics remained informed of all movements of ideas or discoveries? The dictatorship of the one and the insults of the others have changed all that.

After presenting some salient moments in human physical and technical evolution (according to Boule), Mauss then drew conclusions for both the political classes and the working masses:

> How everything becomes relative before such evidence! What teachings for those who believe that modern humans and contemporary societies are the perfect pinnacle of evolution! What a lesson as well for those who, in their haste, imagine that their violence will engender a perfect society, and a race which will need to progress no more. (Mauss 1997 (1921):397, 401)

Whether set in simplified form for a broader popular audience, or expressed in more learned language across his scientific writings, this combined lesson in humanity and humility was undoubtedly salutary for its time and place. By confirming that both techniques and their study are fraught with challenges and implications, this lesson gives us one more reason for reading and appreciating Marcel Mauss today.

Notes

1. This introductory essay is partly based on some of my previous publications, especially Schlanger (1998) for parts III and IV here. There have actually been very few sustained assessments of Mauss's technological ideas, besides frequent and at times obligatory mentions of his classic 'Techniques of the Body'. More developed appraisals are due to 'cultural technologists', members of the *Techniques & Culture* research group in Paris, and to historians and anthropologists of techniques (see, for example, Haudricourt 1987, Lemonnier 1992, Sigaut 1994, as well as Warnier 1999). A recent publication on the topic, by Vatin (2004), came too late to my attention to be taken on board here. Given the aims and scope of this present book, other opportunities will have to be found for assessing the posterity of Mauss's technological principles and insights in the French research tradition, and indeed for appraising the impact and resonance of his work across contemporary social sciences, anthropology, archaeology and material culture studies-for example through the notions of totality, of *habitus* and of traditional efficient acts, or through his incipient sociology of practical knowledge.
2. A further relevant term is of course 'material culture'. Without undertaking here a critical history of this concept and its uses, suffice to note that it was rarely ever used in France until well after the Second World War, in part because of substantive and ideological tensions between the notions of 'culture' and of 'civilisation'. This is obviously not to say that things which count nowadays as 'material culture' were ignored in the French social sciences – on the contrary, we need only mention now (and appreciate below) Durkheim's influential totemic conception, or Mauss's own integrated outlook on processes of production ('technology' in a narrow sense) and of consumption, and more generally his insightful appreciation of the social and symbolic dimensions of everyday objects and practices.

3. 'Driving stone wedges to remove pieces to make a shield, heliman'. P.159.ACH1. New South Wales, Australia. Mounted Haddon Collection (CUMAA©). 'The shield usually is a personal weapon that cannot be lent. In a society that is even moderately warlike, the shield's decoration can indicate the exact rank of its owner' (Mauss, text 10).

Text 1

Technology (1901)

Emile Durkheim

E. Durkheim, 1901, 'Technologie', *Année sociologique* 4:593–94.

Translated by Nathan Schlanger

Durkheim wrote this brief introductory text on technology in the *Année sociologique* as part of his broader attempts to map and to occupy the sociological domain. He then left responsibility for the rubric itself to Henri Hubert (see text 2). Although this introduction ostensibly served to single out a new field of interest, it actually assigned to techniques a rather limited methodological role, as a marker of the state and identity of the civilisation that produced them. Indeed, Durkheim had by that time relegated the technical and material substrate to a marginal position in his sociological explanation, and begun focusing instead on religion.

The various instruments used by humans (tools, weapons, clothing, utensils of all sorts, etc.) are products of collective activities. They are always symptomatic of a determined state of civilisation, such that there are well-defined relations between them and the nature of the society that employs them. The determination of these relations constitutes therefore a sociological problem and technology, considered in this aspect, is a branch of sociology. It is as such that it figures here [in the *Année sociologique*]. Since this science is still but a desideratum, we have in no way sought in what follows to be complete and to assemble all the materials that could serve for this kind of investigation. We have limited ourselves to assembling some writings that seem to us particularly appropriate for drawing the attention of sociologists to these questions.

We include under this rubric matters pertaining to the house, for the house is, after all, an instrument of human life. Up till now, we have included studies relating to habitation in 'morphology', because the form of houses contributes to determine the material form of the groups that inhabit it. But it may be more rational to classify them here; there are so many links between the house and the daily instruments of life.

Text 2

Technology. Introduction (1903)

Henri Hubert

H. Hubert, 1903, 'Technologie. Introduction', *Année sociologique* 6:567–68.

Translated by Nathan Schlanger

Of the *Année sociologique* group, the archaeologist and museum curator Henri Hubert was the most competent to oversee the rubric 'technologie', a responsibility he shared with Mauss from volume 8 onwards. Rarely allocated more than three or four pages per issue, this rubric mostly listed titles of recent publications, and reviewed some works in ethnography (notably German and North American), in prehistoric archaeology and in technological studies. This low-key rubric was reactivated in the *Année sociologique* third series (1948) through the contrasting perspectives of Georges Friedmann and André Leroi-Gourhan. In the text translated here, Hubert proposed that techniques be considered in resolutely sociological terms, as a veritable institution which implicates collective representations in addition to purely mechanical considerations.

We continue to leave this rubric open-ended, without any pretension to be exhaustive. If we quote somewhat specialised articles, on the pottery of Kabylia, say, or on musical instruments, this is because they complete studies already indicated in the *Année*, or because they initiate studies that should, we believe, be of interest to us. Thus, those aspects of the [technical] phenomenon that appears to be regular and apparently necessary, such as the succession of ages and the superposition of industries, could lend themselves to sociological study. Our attention was drawn this year to the question of the invention of forms. Invention does not simply solve a mechanical problem. Between the problem and the solution are wedged a whole series of trials, not to forget extraneous data: it is on all these that sociological research should be carried out. Truth to say, the actual invention of forms usually escapes our observation. To the contrary, we can easily appreciate the insistent conservation of already existing forms. But from a sociological point of view, the invention of forms and their conservation appear to be a single phenomenon, such that the latter can inform us on the former. Invention and conservation should both equally well express the processes of imagination of humans in society. In sum, there occurs in technology what we observe in aesthetics; types constitute themselves, and types of tools, just like types of works of art, are social things and veritable institutions.

Thus are introduced among the given of the problem some elements which do not pertain to mechanics, but rather elements which are obscure, unconscious, and which derive from everything that characterises a group, and within an individual. The whole system of representations of the group is implicated here. This fact clearly transpires when we compare the notion of tool or of machine in our scientific age with that of the times when weapons and instruments were endowed with a soul. It is noteworthy that in general the particularities of the object grant it some individuality, such as the individuality that, even nowadays, is given by soldiers to their gun.

Tools, weapons and all objects of technology are the product of various social things. The relationship that prevails between instruments and societies is a general problem of sociology, which we have addressed elsewhere. Here we need only consider its earliest manifestations.

Text 3

Note on the Concept of Civilisation (1913)

Emile Durkheim and Marcel Mauss

E. Durkheim and M. Mauss, 1913, 'Note sur la notion de civilisation', *Année sociologique* 12:46–50.

Translated by William Jeffrey (amended)

Along with the 1903 essay on primitive classification, this is one of the few texts co-authored by Durkheim and Mauss. At issue here was not only the concept of civilisation per se, but also questions of disciplinary scope and competence. Having argued that the clear-cut and natural human group is the society (and specifically the political society), the authors acknowledged the existence of supra-social phenomena and claimed their sociological importance. Conversely, by addressing other disciplines dealing with these topics (i.e. ethnography and prehistory) as well as other research traditions – and specifically the ascending German geographical and culture-historical sciences – Durkheim and Mauss could argue that it was necessary to take on board their own version of sociology in any investigation of civilisations.

One of the rules we follow in the *Année* is that while we study social phenomena in and for themselves, we also take care not to leave them suspended in thin air, but always relate them to a specific substratum, that is to say, a human group which occupies a determinate and geographically representable portion of space. Of all these human groups, however, the largest – the one which includes within itself all the others and, in consequence, surrounds and envelops all forms of social activity – appears to be the one formed by the political society; the tribe, the people, the nation, the city, the modern state, etc. It seems then at first glance that collective life can develop only within political organisms which have settled outlines and clearly marked boundaries. It would therefore appear that national life is the highest form of collective life, and that sociology cannot know any social phenomena of a higher order.

There exist nevertheless some social phenomena that do not have such clear-cut boundaries; they reach beyond political frontiers and have a spatial expansion that is more difficult to determine. Although their study is at present hindered by their complexity, it remains important to establish their existence and to mark their position within the field of sociology as a whole.

Ethnography and prehistory in particular have drawn attention to this aspect. The immense labour undertaken over the past thirty years in the ethnographic museums of America and Germany, as well as in the prehistoric museums of France and particularly of Sweden, did indeed bring with it significant theoretical results. Especially with regards to ethnography, the scientific requirements of simplification and cataloguing, and even the sheer practical necessities of ordering and exhibition, have led to the creation of classifications that are at once logical, geographical and chronological. These classifications are logical because in the absence of any possible history, logic is the only means by which to perceive, at least hypothetically, the historical sequences of tools, styles, etc. They are also chronological and geographical insofar as these series have developed in time as well as in space, while spreading to a range of different peoples. It has long been the practice in American museums to display distribution maps showing the extent of some particular type of art; and likewise prehistoric museums offer genealogical schemes for the forms of such or such implement.

There thus exist social phenomena that are not strictly attached to a determined social organism. They extend beyond the territory of any single nation, or they develop over periods of time exceeding the history of any single society. They lead a life which is in some ways supra-national.

Technology and aesthetics are not the only fields that present such problems. Linguistics too has long recognised numerous phenomena of the same kind. The languages spoken by different peoples may have many elements of affinity with each other. Some verbal or grammatical forms, found in different societies, allow the grouping of these societies into families of peoples which are or have been related to each other, or which have a common origin: thus, we commonly refer to the Indo-European languages. The same goes for institutions. The various Algonquin or Iroquois nations had the same kind of totemism, the same form of magic or religion. The same type of political organisation (power of chieftains) can be found among all Polynesian peoples. The origins of the family have been identical among all the peoples speaking an Indo-European language.

What is more, it has been recognised that phenomena showing this degree of extension are not independent of one another; they are generally linked in an integrated system. It even happens quite frequently that one phenomenon implies the others, and reveals their existence. Matrimonial classes are characteristic of an entire pattern of beliefs and practices to be found throughout native Australia. The absence of pottery is one of the distinctive traits of Polynesian industry. Some forms of adzes are essentially Melanesian. All the peoples who speak an Indo-European language have a common substratum of ideas and institutions. The foregoing are not simply isolated occurrences, but complex and integrated systems which, without being limited to a determined political organism, can nevertheless be situated in time and space.

To these systems of facts, with their unity and their specific mode of existence, a special name should be given; the most appropriate seems to us to be that of 'civilisation'. No doubt, every civilisation is susceptible to become more national in nature, and to take on some distinctive characters within each people or each state. But the most essential elements which constitute each civilisation are not the properties of a state or a people. They rather overflow frontiers, either by spreading from specific centres by their own powers of expansion, or as a result of the relationships established between different societies, in which case they become their common production. There exists a Christian civilisation which, despite having various centres, has nonetheless been elaborated by all Christian groups. There was a Mediterranean civilisation, common to all the populations bordering the Mediterranean coast. There is a civilisation of Northwest America, common to the Tlingit, the Tsimshian, and the Haida, even though they speak various languages, have different customs, etc. A civilisation constitutes thus a kind of moral milieu within which are immersed a certain number of nations, and of which each national culture is but a particular form.

It is noteworthy that these very general phenomena were actually the first to attract the attention of sociologists, and that they served as materials for the development of this discipline. Auguste Comte was not concerned with particular societies, nations or states; he studied the general march of civilisation. He set aside considerations of national characteristics; they were of interest to him only insofar as they could help him trace the successive stages of human progress. We have often had the opportunity to show how this method deals inadequately with the facts. It leaves aside the concrete reality which the observer can best and most directly attain, namely, the social organisms, the great collective personalities which have been formed in the course of history. It is with them that the sociologist should begin; describe them, arrange them in species and genera, analyse them, and seek to explain the elements of which they are composed. One may even suggest that this human milieu, this integral humanity whose science Comte intended to establish, is scarcely more than a construct of the imagination.

That said, it remains that over and above national groups there exist others, broader, less clearly defined entities, which nevertheless have an individuality and are the locus of a new kind of societal life. While there does not exist *a* single human civilisation, there have always been a diversity of civilisations which dominate and surround the collective life specific to each people. There is here a whole range of phenomena which deserve to be studied through dedicated procedures.

Upon this, all kinds of problems which have hitherto been neglected can now be addressed. One may inquire into the causes of variations in areas of civilisation, why they have halted here or there, what forms they have taken and what factors have determined these forms. All the questions which, as Friedrich Ratzel* has shown, can be raised regarding political frontiers, may equally apply to these ideal frontiers. Furthermore, not all societal facts are equally suitable to become international in nature. Political and legal institutions as well as phenomena of social morphology form part of the specific make-up of each people. In contrast, myths, tales, money, commerce, fine arts, techniques, tools, languages, words, scientific knowledge, literary forms and ideals – all these travel and are borrowed, and therefore result from a history broader than that of a single society. There is therefore room to ask on what does this unequal coefficient of expansion and internationalisation actually depend. These differences do not simply stem from the intrinsic nature of social facts, but also from the diverse conditions in which societies can be found. Thus, according to circumstances, the same form of collective life is or is not amenable to be internationalised. Christianity is essentially international but there exist also religions that are narrowly national. Some languages have expanded over vast territories, but others serve to distinguish nationalities, as is the case with the languages spoken by the leading European peoples.

All of these problems are properly sociological. No doubt, they can be approached only if other problems, which do not pertain to sociology, have first been resolved. It is for ethnology and history to trace these areas of civilisation, to link various civilisations to their fundamental source. But once these preliminary tasks are sufficiently advanced, it becomes possible to launch other general inquiries with sociological bearing, such as those indicated above. What is at stake here is to reach causes and laws by means of systematic comparisons. Thus, we have difficulty in understanding how some writers, notably Father Wilhelm Schmidt, have sought to withdraw the study of civilisations from the field of sociology and reserve it for other disciplines, notably ethnography. For one, ethnography has proved insufficient for the task, and history has to make the same kind of researches concerning historical peoples. Moreover, any civilisation is only the expression of a special kind of collective life, one which has for its substratum a plurality of related and interacting political entities. International life is but a social life of a superior kind, which sociology should take on board. The idea of excluding sociology from these inquiries would not have occurred, were it not still too often believed that the explanation of a civilisation merely resides in searching whence it came from, from whom it borrowed, and by what routes it passed from one point to another. In reality, the true way to understand civilisation is to find the causes from which it has resulted, that is to say, the collective interactions of diverse orders of which it is the product.

4. 'Building a cone-shaped camp'. P.142.ACH1. New South Wales, Australia. Mounted Haddon Collection (CUMAA©). 'The covering [of the conical tent] varies according to regions: in Siberia, deer skins give way further south to strips of birch bark, and then to pine and larch' (Mauss, text 10).

Text 4

The Nation (1920/1953, extracts)

M. Mauss, 1953, 'La nation', *Année sociologique* (third series) 3: 7–68. (Extract 1, pp. 37–41; Extract 2, pp. 49–54.)

Translated by Dominique Lussier

In the aftermath of the First World War, the Treaty of Versailles and the Bolshevik revolution, Mauss wrote substantial notes for a comprehensive study on the phenomena of the nation. These notes appeared only posthumously, however, and in incomplete form, in the third series of the *Année sociologique*. Mauss identified national self-consciousness as a key feature, and then severely criticised the self-aggrandising faith of modern nations in their own race, their own language, and their own civilisation – as developed in the first extract translated here. But nations do not live alone, and Mauss discussed at length the increasingly topical issue of 'internationalism'. The second extract, dealing specifically with civilisations and techniques, praises optimistically the importance of borrowings and exchanges for humanity as a whole.

Extract 1

In the third place [following race, and language], a nation believes in its own civilisation, its customs, its industrial arts, its fine arts. It has the fetishism of its literature, of its plastic expressions, its science, its techniques, its morals, its tradition – in a word, of its character. Almost invariably, a nation has the illusion of being the first in the world. It teaches its literature as if none other existed, its science as if it alone collaborated in its elaboration, its techniques as if it has invented them, and its history and morals as if they were the best and the most beautiful. There is here a natural fatuity, caused in part by ignorance and political sophistry, but often also by the requirements of education. Even the smaller nations do not escape this; each nation is akin to these villages of our antiquity and folklore, convinced of their superiority on the neighbouring village, fighting with 'the crazies' from the opposite side. Their public derides foreigners, just as in Monsieur de Pourceaugnac the Parisians mocked the people of the Limousin. These nations inherit the prejudices of the ancient clans and tribes, the parishes and the provinces, because they have become their corresponding social units, and are individualities with a collective character.

It would take too long to describe all the facts which mark this 'nationalisation' of thought and of the arts. We cannot be expected here to summarise well-known themes in the histories of literature, art, industry, customs and law. The nineteenth and the early twentieth centuries have perhaps focused too much on such themes at the expense of human values, and also as a reaction against the humanitarianism and the Masonic cosmopolitanism of the previous centuries or of the progressive classes in the various nations. Theories of literary history, such as Hyppolite Taine's theory of the 'milieu', applied to England and France; or Johann Gottfried Herder's theory of *Volksgeist* applied to the history of German civilisation – all this occurs in the realm of literary criticism. What is happening in the domain of arts and sciences is perhaps less pronounced, but far more serious.

For one thing, a conscious effort is being made to stay in line with an already heavily felt tradition. Millions of imitations, citations, centos, allusions, have frozen literature into often insipid national forms. Rhythms, canons and customs have fixed the forms of dance and mime: academic authorities, conservatories (so aptly named) have kept invention in check. During the Middle Ages and the Renaissance – given the unity of the Church and the universities, and despite the difficulties of communication, the lack of printing and photography, of patents and licences – the evolution of the arts, sciences and ideas had actually had far more unity and logic than the impact on progress which is brought nowadays by the jolts and clashes of ideas and of modes of artistic expression, by mutual isolations, prejudices and national hatreds. Witness in this regard the French cabal against Wagner, and his stupid revenge. Even industrial techniques have been the object of national traditions, appropriations and conflicts. The Portuguese, the Spaniards and the Dutch reserved for themselves their concessions in India just as the Phoenicians withheld the secret of the Cassiterides ['Tin Islands']. During the seventeenth and even the eighteenth centuries, with regards, for example, to the invention of porcelain, industrial secrets were as well kept as military ones. In this respect, in their attempts at finding out the secrets of others and keeping their own, twentieth-century Germans had manners worthy of the glass-makers of the Venetian Republic. The notion that a nation is the owner of its intellectual property and can plunder with impunity that of other nations is so ingrained that it is only very recently, through the Berne conventions (to which not all states have as yet subscribed) that literary, artistic, technical and industrial property, after being slowly acknowledged by national legislatures, are coming under the jurisdiction of private international law.

Even the forms of law and of economic life, even the unbridled exploitation of land or of subject peoples, have sometimes been construed as grounds for national rights. The conviction that what we call civilisation is a national product has entered people's minds to the extent that it has become a foundation for territorial claims. It is almost comic to see some ill-known, ill-studied folkloric elements being invoked during the [Versailles] Peace Conference as evidence that such and such nation should extend here or there, on the grounds that we can still find there such or such shape of house or some bizarre custom.

Together with that, there has been a sustained effort, especially among the nations of Eastern Europe, to go back to popular sources, to folklore, to the origins, real or imagined, of the nation. It is not only language but also ancient traditions that people have tried to recreate and bring back to life, in some cases successfully. This movement started in Scotland, with the notorious story of the Ossian fake – this supposedly recovered element of Gaelic literature. Then came the German romantics and philologists, of which the fairy tales of Grimm and the discovery of the *Edda* were two decisive moments. It was believed that Germanic civilisation itself had been found. Both poetry and music, above all Wagnerian music, strove hard to feed on these origins in order to revive them – sadly, names of the Germanic epics ended up by being given to the trenches which were supposed to protect the routed army. The Finns and the Slavs followed the German example, and likewise the Serbs, Croats and Czechs built for themselves literature of this kind. Russian music is intentionally folkloric: the principles of the famous 'Four' are known. The museums of ethnography, the return to national arts and the successive fashions that have seized on them, all this boils down to the same fact: while it is actually the nation that makes the tradition, some would on the contrary seek to reconstruct the nation on the basis of tradition.

It is at once comic and tragic to observe the developments given in Eastern Europe to the notion of a 'dominant civilisation'. I shall come back to this topic in the conclusion to this chapter since it is of the highest practical importance. But at this stage of the demonstration we must record the fact itself. In the jargon of the diplomats, folklorists and imperialists, German or Slavonic, used by the pan-Germanists or pan-Slavs and others, the notion of 'dominant civilisation' in a heterogeneous society refers to the property of the civilisation of the dominant group to impose itself, and even assert itself as the sole civilisation of the country. For long, it was in the name of this principle that the Habsburgs ruled over the Slavs and the Hungarians; under their sceptre, the Germans in Cisleithania and the Hungarians in Transylvania were tyrannising the Slavs and the Latins. In the Serbian affair, it was the upholding of these false rights at all costs that provided one of the causes and also the main occasion for the Great War to be declared. One result of this war has nevertheless been to make the application of such principles more difficult, if not downright absurd. The fact that one people has been in a position to block all material and moral development in another is no longer an entitlement to rule over this other people, thanks be the Gods and the Fourteen Points [of the Treaty of Versailles]. While it may be true that in eastern Galicia the only element of 'culture' lies with the Poles, and that the Ruthenians or the Ukrainians are but poor peasants, it does not follow that the right should belong with this so-called élite, rather than with the masses. It is no longer the case that the distribution of the people and their lands should rest with the noble landlords and with the Polish jurists and middle class, as well as with the Jews expediently baptised as Poles. Nor is it true that Bessarabia, taken over by the Russians, should remain Russian, or that the Baltikum should become German because of the predominance of Teutonic barons and partly Germanised Jews.

Extract 2

I. Civilisation

The history of civilisation, from the point of view that concerns us, is the history of the circulation between societies of the various goods and achievements of each. We have already said this in our study of the notion of civilisation, when insisting that societies are not to be defined according to their civilisation, and also when discussing the birth of nations: societies are immersed, as it were, in a bath of civilisation. Societies live by borrowing from each other, but they define themselves rather by the refusal of borrowing than by its acceptance. Consider on this subject the remarkable argument put forward by the king of one of the kingdoms of Tcheou China to his advisers and great feudal vassals, who were refusing to wear the Hunnish (Manchu) dress and to ride horses instead of chariots. He painstakingly tried to show them the differences between rites and customs, between the arts and fashion. Politeness, gestures, and even kissing, all these things that are nowadays spreading and being imitated across nations, were precisely among those things that were known and presented to societies, and yet refused by them.

But this is not the place to study the rejection of borrowing – actually a topic for descriptive, historical, or, better still, psychological sociology, since it is a phenomenon more typical of specific societies and more explanatory than the borrowings themselves. It suffices to have shown that communities are more influenced on this point by institutions than by tendencies, which again proves that borrowing is the normal state of affairs since non-borrowing is precisely what distinguishes one society as against another.

Phenomena of borrowing are all of a physiological, dynamic nature, and they fit rather well into the structure of sociology and the classical social sciences; economics, technology, aesthetics, linguistics and law. Listing the principal phenomena of borrowing would be endless. It is known they served as the basis for Gabriel Tarde's unrealistic and lightweight doctrine [of imitation], and his books contain copious descriptions, albeit banal, lacking in historical perspective and, even more so, in logic. I shall merely indicate here some more typical facts, and particularly stress for each major category of social facts the degree of openness characteristic of modern nations, and the ensuing degree of uniformity for the civilisation of today and of tomorrow. Everything that is social and does not pertain to the very constitution of a society can be borrowed from one nation or one society to another. I shall content myself with demonstrating this, since historical developments have particularly enhanced this human character of institutions, of the technical and of aesthetic arts; so much so that it is now possible to speak in terms of a worldwide human civilisation ['*civilisation humaine mondiale*'], to make use of a modern jargon phrase even before its official approval by the Académie française.

We use the word commerce, and the Latins used with greater precision the word *commercium*, to refer not only to economic relations but also to relations of all sorts, exchanges of every kind and reciprocal transactions of any nature between societies. We ought, however, to distinguish between intra-societal and inter-societal commerce. The exchange of services and goods between such diverse groups as clans, tribes, provinces, classes, businesses, families and individuals is the normal state of affairs in internal social life, and in a large measure creates it. However, we must recall that in segmentary societies commerce between clans and tribes is to some extent conceived as commerce between foreign groups. Thus parallel rules, in the Hebrew and Greek laws, admitted marriage and transfer of real estate inside the clan while banning them outside the tribe. Exchanges used to be curiously restricted in their scope, owing either to the small number of things that were being exchanged, or to the small number of people actually participating in these exchanges.

Commerce between societies was less frequent and more difficult, if not almost impossible. We shall see why in the third paragraph. It was nevertheless all the more serious and solemn, especially when imposed by necessity. Things and customs filtered arduously through the watertight barriers that societies erected between themselves, and they did so through fissures which these societies, no less curiously, allowed to remain. It is only recently that nations have overall ceased to be morally and materially closed to each other: far from restricting exchanges, they have on the contrary increased their quantity, opportunity and intensity. In fact societies today are more or less in the same position as that of the two exogamous moieties, amorphous and confronting one another, at the dawn of the most elementary political and familial organisation. They are also in the same position as were, later on, the tribes, the associated towns, and the small states that followed, before they became the possessions of more powerful states. While ancient law did not recognise the *commercium* and *conubium* of non-nationals, modern nations have for centuries already instituted in their private and public law *commercium* and *conubium* with practically all of humankind. Japanese law is the only exception. It is true, however, that in practice justice is denied the Japanese by important nations, such as the Anglo-Saxons. But in theory the rights of the people were, a few years back, quite contrary to that old code which the Jews had to modify following Hadrian's pressure, and which only granted rights to fellow citizens. Morally, mentally, materially, societies have engaged in the most intimate commerce, at any rate between individuals from all backgrounds and origins.

All this makes it possible to speak of commerce outside the framework of economics. Let us nevertheless restrict ourselves to this use of the term, as is now customary, and say that relations between societies have always, and primarily, been of a commercial kind. This statement will cause surprise, perhaps, but I know of no society so lowly and primitive, or so unfathomably ancient, that has been so isolated from the others as to have no commercial relations. The Australians engage in commerce over long distances; some stones or shells, already used as currency, even travel from tribe to tribe. There even exist markets of a sort in Central Australia; Sir Baldwin Spencer, the great observer of the tribes of Central and Northern Australia, has recently described a strange system of inter-tribal exchange involving pricing in relation to funerary rites (Hakoutou). Similar occurrences must have taken place in Europe ever since the earliest neolithic period; amber and crystals were already then in circulation. Pre-Columbian America, even in the remotest times, also had its amulets, its pottery, and its fabric, travelling over long distances. We ought not to imagine all these societies as lacking civilisation, nor civilisation itself to be that young in the world. The Melanesians, for instance, are great navigators and traders, and they have a currency; likewise for the Naias. The effect of civilisation, in the layman's sense of the word, has been to regulate, multiply and universalise commerce; to have it shift from exceptional or ritualised trade to free barter, from barter to purchasing, from purchasing to a market economy; from colonial or national markets to a world market. In fact, national economies have now so opened up that they have all become dependent not only one on another, but also, and this for the first time in history, they have become absolutely dependent on a certain mood of the world market, in particular the trade in precious metals and value standards. I shall come back to this most instructive fact, due in a great measure to the War. We can clearly see now where to situate this fact, at the summit of a curve whose origin coincides with the origins of human societies themselves.

II. Technique

Initially, commerce concerned first and foremost magical and religious objects, currencies and manufactured artefacts rather than consumable goods. Instruments, utensils, arms, etc. thus travelled over very long distances. Workshops producing various categories of points, arrows and axes were operating in prehistoric Europe, in very remote times. With the help of Australian ethnography, we can imagine how such commerce functioned, and the prestige it bestowed, often of a magical kind, upon the tribe possessing both the raw materials and the technique needed to transform them into tools.

This inter-societal commercial dimension of technical life has developed to such an extent that today the unfortunate peasant in Soviet Russia is unable to repair or maintain even the coarsest ploughing tools, all of which were produced abroad, and has as a result to revert progressively to the most primitive form of agriculture.

Yet all this commercial aspect is less important than the phenomenon of borrowing and the spread of techniques. In normal circumstances, and unless it is deprived of manpower, of will or of raw materials, or on the contrary if it is ridden with prejudice, a society will strive to acquire and make its own the techniques it recognises as superior. This is often a necessity, especially in military matters, since superior weapons grant to this society the power of life and death over its neighbours. Most of the time, however, borrowing is a matter of self-interest, a quest for a better and easier life. Economic pressure of this type has only appeared with the development of means of communication on land and on sea, the appearance of money and the formation of a world market. At the risk of undue impoverishment, nations had to keep pace with others and reach the same technical level. But this goal was consciously pursued only from the sixteenth century onwards, during the early development of large-scale commerce and capitalism and despite the extreme protectionist measures of the time. Changes occurred then as a struggle between nations, quarrelling over state secrets with which to acquire riches. We can say that all great industrial civilisations ever since have developed an international character. Industry was undergoing before the First World War a process of standardisation and equivalence, expanding and diversifying throughout the world. Exchanges of techniques were taking place, and with them an intensified exchange of products, the one entailing the other. A worldwide industry was in the process of being created, made possible by the exchange of products and raw materials on a global scale.

The dream of the 'plain of cabbages' so ridiculed by Alfred de Musset had by and large become reality. Against the absurd reservations of the hacks and the nationalists, we cannot overstate the importance of technical borrowings and the human benefits that ensue. The history of human industries is properly the history of civilisation, and vice versa. The diffusion and the discovery of industrial skills have been and continue to be the fundamental progress which permits the evolution of societies – that is to say, the increasing happiness of greater and greater masses over wider and wider territories. It is technique which, through the development of societies, has brought about the development of reason, sensibility and will. It is technique that makes of modern humans the most perfect of animals. It is the Prometheus of ancient tragedy. With it in mind let us reread Aeschylus' magnificent verses and agree that technique has made humanity – feeble ants roving in sunless caves, children who saw not what they saw, heard not what they heard, who throughout their long lives blurred their sensory images with the phantoms of their dreams. It is technique that makes humans equals and that worries the gods; it is undoubtedly technique that will save humanity from the moral and material crisis in which it is struggling. Humanity has been rescued from misery, hazards and stupidity because it domesticated and propagated animals and crops, and progressively transformed the soil itself to suit its needs. Humanity thus became the master of itself and its own destiny. Science and human industry are superior to fate, not subjected to it. Humanity is the third god which puts an end to all gods, to the tyrants of the skies and of the earth. More than land and capital as such, the common heritage of humanity resides in the art of making them grow, and this wealth of products makes humanity possible, a humanity that is internationally civilised.

Text 5

The Divisions of Sociology (1927, extracts)

M. Mauss, 1927, 'Divisions et proportion des divisions de la sociologie' *Année sociologique* (second series) 2:98–176.
(Extract 1, pp. 106–07; Extract 2, pp. 117–23.)

Text published in full in Mauss (2005).

Translated by William Jeffrey (amended)

This is one of Mauss's most programmatic texts, a relaunch of the *Année sociologique* (second series) with an updated survey of the sociological domain in the light of recent theoretical and disciplinary developments. Careful to emphasise continuities with Durkheim, Mauss nonetheless pointed at various problems and omissions. He notably proposed to complement the division between special and general sociologies with one based on the social group ('morphology') and on its representations and actions ('physiology'). Linguistics, aesthetics and especially technology deserved therefore far more space and attention than they had previously received, as Mauss explains in the second extract translated below. In the first extract, he provides some all-too-brief comments on technical practices and modes of thought, thus broadening the reach of the Durkheimian sociology of knowledge.

Extract 1

The production of recent years has been directed towards a third problem, whose philosophical interest has rendered it popular. Thanks to us, it is at present fashionable to interrogate sociology on the *origins of reason*, the primitive forms of thought, etc. Scholars have abstained, wrongly in our opinion, from studying evolved or semi-learned forms. In the *Année*, these problems are unsatisfactorily divided between 'general sociology' and 'religious sociology'. Thanks to Antoine Meillet, we touch on these problems also under 'linguistic sociology'; we shall have to return to this point later. However, to repeat the point, we know that they really belong to that general sociology whose basic outlines Durkheim formulated quite early. However, instead of forming part of the prolegomena, they form the conclusions of our sciences and not merely of some of them, but of the whole set. In this domain, partial considerations are infinitely dangerous. The notion of class or *genre* is mainly juristic in origin, as Durkheim and I have assumed; as Hubert has said, the notion of time, and as Durkheim wrote in the *Elementary Forms of the Religious Life*, the notion of soul and, in some pages of the same book which have been too little noticed, the notion of the Whole are mainly religious or symbolic in origin – none of these arguments mean to say that every other general notion has had the same kind of origin. We do not at all believe that. There remain to be studied many other categories, both living and dead, deriving from many other origins, and in particular categories of a technical nature. To cite only the mathematical concepts of Number and Space, who will ever say enough and with sufficient exactitude the part which weaving, basket-making, carpentry, nautical art, the wheel and the potter's wheel have had in the origins of geometry, arithmetic and mechanics? We shall never tire of recalling the splendid observations by Frank H. Cushing*, profound observer and inspired sociologist, on 'manual concepts' (*Amer. Anthrop.*, vol. 5 (1892); see. *Année*, Vol. II). We would never come to an end of listing the various activities and also the various ideas whose forms are at bottom general ideas, including those which are still at bottom our own ideas. These studies of the forms of thought, primitive or not, should appear at the end, to crown and to synthesise our studies.

Extract 2

Technology. Even though one of us, Henri Hubert, archaeologist and prehistorian, is by profession a technologist, we have never had the necessary time and energy to give to the technical phenomenon the formidable place which it merits.

'Homo faber', says Bergson. These formulae signify only the obvious or they signify too much, because the choice of such a sign hides other equally

time they have taken it out of itself. Here again, *homo* is *duplex*, but he is so in a different sense than in law or religion. In religious ecstasy, in moral sacrifice, around the Golden Calf, the individuals and society always remain themselves with their limits and imperfections. In practical arts, human beings make their limits recede. They advance in nature, at the same time above their proper nature, because they adjust it to nature. The human actor identifies himself with the mechanical, physical and chemical order of things. He creates and at the same time he creates himself; he creates at once his means of living, things purely human, and his thought inscribed in these things. Here is true practical reason being elaborated.

* * *

Technology and natural history of the sciences. Perhaps it is also in techniques and in relation to them that true reason itself is elaborated. We must recognise that the plan proposed and our studies themselves present a very serious gap on this point, the most serious perhaps among those which concern the special parts of sociology. Contrary to the Comtean tradition, we nowhere study for itself the natural and social history of the sciences. It is not that we lack support. The distinguished editors of *Isis: Review of the History of Sciences*, Abel Rey and others, also conceive their work as eminently sociological. Nevertheless, until now, we have not made an effort to assign a place to these studies, still less to estimate their extent and depth. Some observations are therefore necessary here.

When the arts and the sciences and their historical relations are studied concretely, the division between pure reason and practical reason seems scholastic, scarcely realistic, barely psychological, and even less sociological. We know, we see, we feel the profound links which unite them in their raisons d'être and in their history. Particularly strong at the origin, they are still obvious today when, in a thousand cases, techniques pose the questions which science resolves and often create the facts which science mathematise or schematises after the fact. On the other hand, quite often, theoretical discovery proposes the phenomenon, the principle, or the invention which industry then exploits. The scientific-technical complex forms a single bloc. For example, the oldest calendars are as much the work of farmers as of religious minds or of astrologers; technique, science and myth are there blended. In the same way, pigeons had been selectively bred before Darwin found the notion of natural selection. The same is true of pure and experimental science – which in our days replaces mythologies, metaphysics and pure action, even action based on reflection; it is not in the least disengaged from the action which it directs, even when it detaches itself most clearly or most deliberately. Have not the most modern doctrines of cosmology finally led to purely practical researches? Attempts are made to find a stable measure in the single constant known at present: the length of a light wave.

This is why the history of sciences and epistemology perhaps needs to be situated in a special part of the sociology of techniques. In fact, science is the other social activity which, like techniques, draws humans out of themselves towards nature, inspires their technical activities and has the same goal, namely control over things.

Nevertheless, we hesitate before this radical solution. This arrangement neglects a specific difference. In their industrial arts, humans remain humans and only half emerge from themselves. In contrast, science makes humans emerge completely, and identify with things. Once engaged in science, the human being is aware of things in and for themselves, rather than feeling them exclusively in relation to himself and to his acts, or of representing them in a kind of magic mirror, in relation to sometimes useless mythical images.

And from this there arises a second difference between sciences and techniques. However expansive and imitable techniques may be, they are, even in our day, relatively variable between nations. In contrast, while science remains social in so far as it is due to collaboration and controlled verification by humans, it nevertheless ceases to be the work of societies as such. More and more it is the treasure of the entire human community and no longer of any particular society. Having formerly been fashioned from jealous traditions, secrets and mysteries, alchemies and recipes, it is now practised in broad daylight and belonging to humanity. In order to approach its study, perhaps it is necessary to abandon at the outset the partial viewpoint of past or present societies, and rather adopt the viewpoint of the greatest possible society – humanity. For these last two reasons it may be necessary to add epistemology as a new division to sociology.

On the other hand, perhaps it is preferable to leave science to its natural connection: geometrical, mechanical, physical and chemical practical activity on things, and also rational practice on animate creatures and humans, the agricultural, veterinarian and medical arts. Indeed, it might be better, like Alfred Espinas and the Greeks he followed, not to distinguish *techné* and *epistémé*. Perhaps, like Durkheim, one should separate them profoundly without contrasting them. We could balance the pros and the cons endlessly; we open the debate and we do not know how to close it. Like the good Pindar, we do not know what is just.

From the place which we assign to them, however, one can see how far the problems of science and techniques are fundamental, and condition the problem of the social origins of reason. And, let it be said in passing, here is an additional motive for putting this branch at the end and not at the beginning of our studies.

Text 6

Debate on the Origins of Human Technology (1929)

M. Mauss, 1929, 'Débat sur l'origine de la technologie humaine', *L'Anthropologie* 39:129–30.

Translated by J.R. Redding

At the Muséum d'histoire naturelle and the Institut Pasteur, Paul Guillaume and Ignace Meyerson* had been conducting experiments on monkey tool-use and intelligence, using objects, diversions and intermediary tests. Guillaume was mostly interested in the Gestalt psychology, and Meyerson aimed to identify thresholds and ceilings in the uses of tools by primates, and thus grasp the 'entry' into the properly human. Commenting on their film and presentation, Mauss reproduced the commonplace view on the specificity of human technicity – a view that was subsequently to be revised by his student Leroi-Gourhan – but he also pointed at the need to assess technical capacities in 'real life' settings rather than through artificial experiments.

Guillaume and Meyerson presented their film on The Psychology of Monkeys with a commentary by Guillaume.

Invited by Meyerson to give an opinion from a technological point of view on the experiments carried out, Mauss declined. He did not believe, however, that they were very significant from the standpoint of a theory of technical creation. The stick used by Nicole was a simple means rather than a tool. Similarly, the set of cords showed the animal's intelligence in a favourable light but did not, in the end, take it beyond the mechanical environment in which it grew up: the rigging of the different cords was somehow too reminiscent of the lianas of the forest. Finally, all the experiments of Guillaume and Meyerson's, like those of Köhler, seemed to consist exclusively of mechanical and principally geometrical problems, to do with scales or parallels, etc. or with boxes with which to cover distances, or again fishing rods requiring adjustment, etc. These experiments prove the importance of 'forms' or rather of 'appearances'.

Mauss is of the opinion that it may be perhaps time to move beyond this type of experiment. He suggested, for example, a study of the capacity to make use of things whose action was physical and mechanical, such as a weight. Might it be possible to verify the famous story of the orang-utan opening oysters or see whether chimpanzees could have the idea of making use of a stone, then employ it to break a box, consider its weight, and then, from stones of the same weight, evaluate its form and, finally, use this stone in relation to its particular form? All this of course without training.

All this could enlighten us on the origins of human technology. For indeed, it must be admitted that there remains a gulf between Nicole's stick or Köhler's boxes and the most elementary pre-Mousterian industry. On the one hand, there are only distances covered or bypassed, on the other, there are objects of definite forms, used in a specific way.

Text 7

Civilisations, Their Elements and Forms (1929/1930)

M. Mauss, 1930, 'Les civilisations. Eléments et formes', in Fondation pour la Science – Centre international de synthèse (eds), *Civilisation. Le mot et l'idée*, 1re Semaine internationale de synthèse, fasc. 2. Paris: La Renaissance du Livre, pp. 81–108.

Translated by J.R. Redding

The 'Semaines internationale de synthèse' were created by the historian Henri Berr as a forum for interdisciplinary research, applying a social sciences approach to current debates. The notion of 'civilisation', the changing meanings and uses of the term, its relations with 'culture' and with 'kultur' was clearly one such topical issue. The conference dedicated to it was coordinated by Lucien Febvre, who was just launching at that time the first issue of the pioneering historical journal *Annales d'histoire économique et sociale*. As Mauss explained in the opening paragraphs of his contribution, he was capitalising here on his extensive readings and reviews for the *Année sociologique* to draw a wide panorama of contemporary 'civilisation studies', including works by ethnographers, historians, philosophers and archaeologists, in French, German and English. This text notably shows that Mauss's approach converged in some aspects with the dominant culture-historical paradigm, and that it also differed, notably in emphasising the social nature of facts of civilisation, and in the positive, relativist evaluation of mixings and convergence.

Introduction

This is an extract from a long methodological note on *La Notion de civilisation*, to appear in volume III of the *Année sociologique*, second series. It was prepared from various notes on the subject: volumes X, XI, and XII of the *Année sociologique*, first series, as well as numerous and lengthy reviews of general works by archaeologists, historians of civilisation, and most particularly ethnologists, published in the two series of the *Année sociologique*. Current advocates of the 'cultural history' method, of 'historical ethnology' and of the principles of 'diffusion', tend to oppose, without justification in my view, their methods and those of the sociologists. We will not engage here in a critical discussion of these theories and of their results. They all count honourable scholars as protagonists. We shall not criticise the supporters of W. Foy and Fritz Graebner any more than those of Father Wilhem Schmidt, or of those of the American School of 'cultural anthropology'. These latter scholars, especially Franz Boas and Clark Wissler, among others, work with societies and civilisations which have obviously been in contact with one another. They are therefore more discriminating than their European colleagues, they generally take care to avoid unruly hypotheses, and here and there they have genuinely been able to identify 'layers of civilisations', 'centres' and 'areas of diffusion'.

These theories are primarily – and all too easily – opposed to the simplistic ideas which represent human evolution as if it were unique. In this respect, just like these 'comparativists', especially historians and geographers, sociologists seek to connect the phenomena of civilisation not to some hypothetical general evolution of humankind, but rather to the chronological and geographical interconnection of societies. Neither Durkheim nor myself have ever separated the evolution of humankind as a whole from the evolution of the smaller or larger groups which compose it. Long ago Durkheim accounted for the modern conjugal family by the mixture of Germanic and Roman domestic laws. Thus, he was applying what is called nowadays the theory of 'substrates'. And it is now more than ten years since Antoine Meillet has opted for a genealogico-historical method in linguistics, without, for that, believing that he was being disloyal to sociology, of which he is a master.

Moreover, all these conflicts between schools of thought are futile intellectual exercises, or competitions between philosophical or theological cathedra. The truly great ethnologists have been as eclectic in their choice of problems as of methods, which must vary according to the problem. E.B. Tylor, who is usually singled out, has actually published – and even better, taught – delightful accounts of borrowings. The best collections informing on the distribution of ethnographic objects are undoubtedly those of the Pitt Rivers Museum in Oxford, which E.B. Tylor founded and which Henry Balfour administers.

In fact, the majority of genuine scholars remain loyal to the three principles or rubrics set out by that old master, Adolf Bastian: I. The *Elementargedanke*, the original and primordial 'elementary idea', an autonomous and characteristic creation of a collective mind, or the 'cultural trait', as the American 'social anthropologists' rather clumsily put it. II. The *Geographische Provinz*, the 'geographic sector', at times quite ill-defined, at others very obviously delimited by shared facts of civilisation, by related languages, and fairly frequently by racial unity: the number of these 'geographic provinces' is not over-large, and modern discoveries are restricting their number even further. III. *Wanderung* – migration, travel and the vicissitudes of civilisation – and, with it, as in the case of autonomous evolution, the *Wandlung* of civilisation, the transformation of civilisation by the borrowing of elements, by migrations, by mixtures of the peoples bearing these elements, or by autonomous activity on the part of these peoples. Let us therefore assume or take for granted this agreement among scholars, and examine how civilisations can be analytically and synthetically studied.

We shall not recapitulate here the history of the word 'civilisation' with its diverse meanings, and nor shall we critique its various usages. The notion of civilisation is certainly less clear than that of 'society' which it actually presupposes. What follows are simply a few definitions which, we believe, will make it easier to know how to talk of these matters.

I. Facts of civilisation

Let us first define what differentiates phenomena of civilisation from other social phenomena. This will enable us to understand what is a system of such phenomena: a civilisation. We shall then see how, from this point of view, one can return without too much difficulty to some fairly wide uses of the term.

The phenomena of civilisation (*civis*, citizen) are by definition social phenomena of given societies. But not all social phenomena are, in the narrow sense of the term, phenomena of civilisation. There are indeed some phenomena which are entirely specific to the society in question, which differentiate it and set it apart. They are usually to be found in the dialect, in the constitution, in the religious or aesthetic custom, in fashion. China behind its wall, the Brahman inside his caste, the people of Jerusalem in relation to those of Judea, those of Judea in relation to the rest of the Hebrews, the Hebrews and their descendants the Jews in relation to the other Semites – all these set themselves apart in order to concentrate on themselves, in order to separate themselves from the others. These examples show that it is better not to talk of civilisations when speaking of phenomena which are limited to a given society, and that it is better in these cases to say quite simply 'society'.

But even in the most isolated societies there exists a whole body of social phenomena which must be studied separately and in their own right, if one is to avoid errors or, more precisely, unwarranted abstractions. These phenomena all share one important feature: they are common to a larger or smaller number of societies and to a longer or shorter period in the past of these societies. They can be labelled 'phenomena of civilisation'.

Social phenomena can thus be readily divided into two main groups, whose size cannot be fixed a priori and whose relative importance varies according to time and place. The phenomena in the former group are unfit to travel, while in the other they are naturally able to do so: almost by themselves they overflow the (often difficult to determine) boundaries of a given society.

All techniques could be borrowed, if they were wanted or needed, and if the means were available. In reality, generally speaking and with some exceptions, techniques are always transmitted from group to group, from generation to generation. Likewise, some aspects of the fine arts could easily be propagated, even the musical and mimetic arts, and this, even in populations as primitive as the Australians. This is the case with what is called there in local English the *corroboree* (the word is of Australian origin) – masterpieces of dramatic, musical and plastic art, a sort of series of great tribal dances, sometimes setting hundreds of dancer-actors in motion, with choirs formed of entire tribes. These *corroboree* are passed from tribe to tribe, are given forever, like an object, like an item of property, a piece of merchandise, a service, like ... a cult, or a magic recipe. The Black African orchestras are constantly on the move in wide areas; the witch-doctor cum minstrel and soothsayers range over even greater distances. Tales are repeated very far afield, over long periods of time, faithfully reproduced in all the directions they reach. The coinage – cowries in Africa, shells in Melanesia (*conus millepunctatus*), *haliotis* mother-of-pearl in northwest America, brass wire in equatorial and central Africa – are all really international: some even have exchange rates. Since the Middle Palaeolithic period in Europe, amber, quartz and obsidian have been the object of intensive and long-distance transactions.

Even those phenomena which seem the most private to the life of societies – secret societies or mysteries, for example – are subject to propagation. The story of the 'Dance of the Serpent' in North America is well known, as is that of the 'Dance of the Sun' over the whole expanse of the Prairies. Henri Hubert and myself have drawn attention in several reviews to these special cults, more or less attached to local bases, through which, as in antiquity, scores of religious, moral and scientific ideas have been transmitted across many so-called savage or barbarous societies.

Even institutions can be borrowed, even the principles of social organisation can gain hold. Thus the concept of constitution, πολιτεια, born in the Ionian world, is propagated throughout Hellas, receives a philosophical formulation, arrives in Rome, *res publica*, then in our civilisations where, after having persisted in urban constitutions and charters, as well as in the small rural and mountain republics, it reappears in state constitutions. One can take up the curious history of the word 'tribe', a word which in Greek and Latin means three and which sometimes rather designates organisations in twos, fours, etc. Military institutions have necessarily been borrowed, up to the present day, just like the techniques of armaments which depend on them and on which they depend. A given fact can thus impose itself beyond the society and the time of its creation.

These *phenomena of civilisation* are thus essentially international, extranational. They can therefore be defined, in opposition to the social phenomena which are specific to such and such society, as *those social phenomena which are common to several societies, more or less related to each other*; be it through prolonged contacts, through some permanent intermediaries, or through relationships from common descent.

A phenomenon of civilisation is therefore by definition and by nature a phenomenon spread over a greater mass of population than the tribe, clan, small kingdom, or confederation of tribes. Thus the features of the Iroquois civilisation are common to all the Iroquois nations, well beyond the league of the Five Nations.

It follows from all the above that the study of such phenomena of civilisation can have a geographical and historical interest as well as a sociological one. Indeed, these facts of civilisation always extend in space, and have a more extensive geography than the political geography of each given society; they cover a wider area than the nation. Moreover, like all other social phenomena, they have a basis in the past, in history; but since this historic past is not that of a single nation, and since it always covers quite large stretches of time, it can be inferred that these facts will demonstrate connections that are at the same time historical and geographical. It is always possible to infer on their basis a fairly large number of direct or indirect contacts, and even occasionally to trace lineages with some certainty.

The observation of these phenomena of civilisation, when accompanied with the observation of other historical and geographical facts, makes it possible to support hypotheses regarding the extension and history of civilisations and of peoples. This in turn allows us to establish a genealogy of facts, of more or less certain sequences, without which neither history nor human evolution can be conceived.

This is then where we situate the observation and study of borrowings, the study of historical filiations, of techniques, of arts and of institutions. Upon these, we can conjecture or asses the occurrence of simultaneous evolutions starting from common principles, or the occurrence of more or less contingent transmissions, but always dominated by the existence of predetermined connections between given societies. On the subject of borrowings can be recommended Eisenstädter's good if already old dissertation, *Kriterium der Aneignung*, in Buschan's collection of *Hefte*. Erland Nordenskiöld's* studies in South America are models of the kind. I myself, together with Nordenskiöld, have encouraged the works of Alfred Métraux on the elements of the civilisation of the Tupis (which incidentally includes numerous elements common to the Tupis and the Caribbeans).

The study of these spreads of elements of civilisation is often extremely curious. It seems difficult to deduce all that Graebner does from the distribution of the sculptured crouching figure (*Hockerfigur*), but the facts he discovered are incontestable. I do not think that M. Jackson was right to interpret the very widespread use of the conch by postulating its Egyptian origin, à la Grafton Elliot Smith. Nevertheless this use is a very clear and important fact of civilisation, and not of simultaneous evolution.

It is indeed against the background of international phenomena that all societies stand out. It is on the basis of civilisations that societies develop their distinctive features, their idiosyncrasies and their individual characters. It should further be recorded how much these traits of civilisations can remain deep-seated and uniform, even after prolonged separation. Thus for example the pygmies of the Andamans have remained the purest, in their islands, with their language, the only known pygmy language. The civilisations of the Gulf of Bengal have scarcely touched them, despite ongoing relations over several millennia. And yet, the pygmies of Malacca, to mention only them, who seem to have lost their language and who live in a Malaysian and non-khmer environment, have largely the same material civilisation as their Andaman brothers.

II. Civilisations. Forms of civilisation

But besides the elements of civilisations, civilisations themselves have their own individual traits and arrested forms, and their conflicts with one another. This is precisely what characterises civilisations: these borrowings, these common features, these coincidences; but also the end of these contacts, the limits of these coincidences, the actual refusal of such contacts with other civilisations.

We may therefore propose the following definition of *a civilisation*: *it is an ensemble of phenomena of civilisation that are sufficiently large, sufficiently numerous, sufficiently important in both quality and quantity. It is also a fairly large ensemble of societies which present these phenomena. In other words, a sufficiently large and characteristic ensemble to be able to signify and evoke a family of societies.* A family of societies which one has other factual grounds to establish. These facts are both current and historical in nature, as well as linguistic, archaeological and anthropological; these facts which lead one to think that they have been in prolonged contact, or that they are related to each other. An ensemble of facts, an ensemble of features of these facts corresponding to an ensemble of societies, in a word a sort of hyper-social system of social systems, that is what can be called a civilisation.

It is consequently possible to talk about more or less vast or more or less confined civilisations. One can even distinguish layers, concentric spheres, etc. For my part I have long been teaching the possibility of a very ancient civilisation all through the coasts and islands of the Pacific. Within this very extensive, rather diffuse civilisation, we can and probably should single out a civilisation of the South and Central Pacific. Within the latter, a Malayo-Polynesian civilisation, a Polynesian civilisation, a Melanesian civilisation, and a Micronesian civilisation are clearly perceptible. It is even permissible to build up all sorts of constructions on the relationship of these four civilisations to each other; and even on their connections with an Austronesian, an Austro-Asiatic and a pan-Asiatic civilisation. For indeed, there are within this immense domain, a large number of coincidences and a large number of variations among these civilisations. Some of these allow us to believe in the original unity of the civilisations, even when there is at least partial diversity of races: for example, black Melanesian and light yellow Polynesian. Conversely, these coincidences can lead us to believe in diversity where there is relative unity of language, for example, Melaneso-Polynesian (we are leaving out the Papuan element). The limits of extension of the betel and the kava, of the bow and the sabre, of the breast-plate and the palisade, of the house on piles, etc. all allow us to classify civilisations and even to advance hypotheses on their genealogy, in the same way that dialectological divergences and resemblances are one of the best means of establishing the family relations of peoples.

It follows from all this that every civilisation has both an area and a form. In fact it always has its halting points, its limits, its core and its periphery. The description and definition of these areas is of prime importance for the history and consequently for the science of humankind. But this area of extension is perceived only because one has the impression that the elements or the phenomena, which together compose such and such civilisation, have a distinctive type or form of their own. The definition of this type is therefore essential, all the more so because the two terms – area and form – are reciprocally connected. Every civilisation has an area because it has a form, and this form can only be perceived because it is spread out over this area and nowhere else. Despite being a second-degree social phenomenon, a civilisation, like every society, has its frontiers and its spirit. The definition of the area of a civilisation therefore calls upon its form, and inversely the definition of a form builds on the area of its extension.

Let us define these two terms. The *form of a civilisation is the sum* (the Σ) *of the specific aspects taken by the ideas, practices and products, which are more or less common to a number of given societies,* who invent and bear this civilisation. We could also say that *the form of a civilisation, is everything which gives a special aspect,* unlike any other, *to the societies* which compose this civilisation.

An area of a civilisation is the geographic extent of the distribution of the total (more or less complete in every society in this area) *of the common phenomena* regarded as *characteristic,* as typical of this civilisation. The area of a civilisation is also the ensemble of land surfaces inhabited by societies sharing the representations, practices and products which compose the common heritage of this civilisation.

By abstraction, and in order to meet the requirements of a short didactic exposé, we will not follow here the current fashion of ethnological science and historical geography and give any consideration to the notion of 'layers of civilisation'. This notion remains, however, very important in so far as it corresponds to what historians call, not very accurately, style, period, epoch, etc. Here is nonetheless a provisional definition: we shall call *a layer of civilisation the form which a given civilisation* of a given extension *takes in a given time.* Such are the main divisions of the facts and of the problems.

* * *

These notions of form and area of civilisation have been used with some exaggeration by two opposing German ethnological schools. The members of one school make of the area of civilisations a means for tracing genealogies, while the others, for the same purpose, use the forms of civilisations. Let us see where both go wrong.

Among the first school, W. Foy, Fritz Graebner, Father Wilhelm Schmidt and his followers take as their starting point the notion of area of civilisation (*Kulturkreise*) and layers of civilisation (*Kulturschichten*). Each civilisation is defined by one dominant feature, and it is almost exclusively the geographical extent and occasionally chronology of this feature that is studied. They talk of *Bogenkultur, Zweiklassenkultur, freivaterrechtliche Kultur,* about the culture of the bow, of a two-class culture (societies divided into two matrimonial halves), of civilisations with male descent without exogamy. And they end up with absurdities, even verbal ones, such as that of the 'totemic axe'. It may be that on a number of points of detail, these authors have actually found probable relationships which are interesting and worthy of historical recognition. But what is valid for studying the distribution of objects easily becomes inaccurate when it is the definition of civilisations and the contacts between them that is at stake. The method of distribution maps is excellent when it comes to describing the history of each instrument or type of instrument, art, etc. It furthermore stands on good grounds when tangible objects are concerned, and when the aim is to classify these objects into series in museums. In this narrow sense, the method meets with our complete approval. From this perspective, for example, a great deal could be said about the greasy pole. The late lamented Robert Hertz* had prepared some good work on the distribution of kites across Polynesia. But to seek to trace the spread of an art or an institution, and to define a whole culture, are quite different matters. Here two dangers immediately emerge.

First, the choice of the dominant characteristic. The biological sciences suffer enough from this notion of principal characteristic, which in our opinion is entirely arbitrary. The sociological sciences suffer from it even more. The criteria employed to identify these characteristics are often quite non-existent. For example, the idea which gives its name to the *Zweiklassenkultur* corresponds to a serious error. It is in reality a purely imaginary assertion that certain Australian and Melanesian societies are divided *only* into 'two exogamous classes' (Graebner and Schmidt), into two 'moieties' (Rivers' terminology). First, we vehemently reject the demarcation established between these moieties and the clans; in our opinion, these moieties are ancient primary clans. Next, in *all* the Australian and Melanesian societies regarded as representative of this *Zweiklassenkultur civilisation*, things other than these moieties have been found: clans have also been observed among them, a normal state of affairs in what we call in our terminology, phratries. It is therefore an error and a petitio principii to separate the fact 'class' from the fact 'clan'.

As well, it is never clarified what are the relationships between this chosen dominant characteristic and the other characteristics of a civilisation. It is not proven that the one necessarily entails the others, so that where the bow is found, for example, we would be more likely to find uterine, or on the contrary masculine, descent (the principle actually varies with each author). This inevitability in the simultaneous distribution of the simultaneous elements of civilisation is by no means demonstrated. Such a demarcation of a layer or an area of civilisation can often result in other absurdities. Oswald Menghin, for example, goes so far as to speak of 'uterine culture' in respect to the Palaeolithic of the Congo. It is admirable indeed that anyone should be able to reconstruct the legal system of an unknown population on the basis of a few worked stones. All this is but fiction and hypothesis.

A civilisation is defined not by one characteristic, but rather by a certain number – generally quite large – of them, and even more so by the respective doses of these different characteristics. For example, navigation holds naturally a different place among the maritime Malayans and Polynesians than it does among the other continental Australonesians. Let us then conclude that the *Kulturkreise* method is poorly handled. This is mainly because it is practised in isolation from the following method.

The second method for the study of civilisations comes with the somewhat pretentious name of *Morphologie der Kultur*. It is primarily known through the names of two authors, as controversial as they are popular: Leo Frobenius and Oswald Spengler. According to Frobenius, maps showing the distribution of all manner of things would make it possible to detect – particularly in respect of Africa – the various cultures and even the roots of culture of which African civilisation is composed. The current civilisations of Africa are almost all intermixed. But from among the combinations and stratifications, Frobenius claims the ability to retrieve those pure cultures whose form is fixed, and whose material utility, moral value and historical importance can be appreciated by the eye of the morphologist. Thus one would see in black West Africa, grafted one on top of the other, five or six civilisations apparently well known to Frobenius: the Aegean and the Syrtic and the South Eritrean and the 'Ethiopian tellurism' and lastly, of course, Atlantis with its 'primitive Eros'. The only serious thing this school has produced are descriptive card indexes, which appear conscientiously done and are useful; this is the beginning of an *Atlas Africanus*, some parts of which are actually valuable.

Equally literary, in our opinion, is Spengler's *Morphologie de la civilisation*. These moral classifications of civilisations and nations into hard and soft, into organic and loose, together with this philosophy of history with its vast and colossal considerations, are really of value only to the general public. It is a regression devoid of any precision into the antiquated formulae of 'cultural destinies', 'historic missions', into the whole jargon of this unconscious sociology that encumbers vulgar history and reaches even the self-styled social science of political parties. The sociologist would truly find more ideas and facts in Guizot. However, we will await an 'Historical Atlas of Civilisation' in order to judge the heuristic value of a not inconsiderable work, whose usefulness is indisputable. Nevertheless, we also have our fears about this work. If here again morphology must be separated from the simple cartography of areas and layers of distribution of objects, etc., if it is guided by the a priori ideas of '*the* culture' or of a priori defined 'such and such cultures', this work will only be full of petitions of principle.

At the end of the day, one can see that these methods, these notions, are legitimate only in so far as they are all used together. To conclude this discussion of ethnographic procedures, it must again be added that we do not consider them to be of a very high level of certainty. They are useful, but rarely sufficient in their own right. To reconstruct the history of peoples allegedly without history by using the hypothetical history of their civilisation is a daring undertaking indeed. We can clearly admit that, on this specific issue of the history of peoples, ethnographic and sociological notions are actually less reliable auxiliaries to the more precise results of linguistics and archaeology. When they are used together, however, they can lead to noteworthy results. Consider in this light the work of the American 'social anthropologists'. The hypotheses advanced by Boas regarding the mythology distributed in the northern basin of the Pacific are more than probable, they are almost conclusive. The hypotheses of Wissler on the Asiatic form of North American Indian clothing are obvious, as are Bruno Adler's older hypotheses concerning the north Asiatic arrow in America. But their proof has been definitely established only with William Sapir's linguistic discoveries, which related considerable groups of North American languages to a proto-Sino-Tibeto-Burmese root.

* * *

The historic uncertainty that will inevitably occur in specific cases should not, however, discourage research: the general facts remain. On the one hand, it is undeniable that civilisations exist, and that they characterise families of peoples or layers of humankind, or both at once. It is equally certain, on the other hand, that each civilisation has its 'aspect', and that their products have their style, their *facies*, that can be analysed – provided this analysis is not made through a single dominant characteristic but by implicating all of them. And all these characteristics have one common trait which requires us to take them into consideration: the trait which constitutes their arbitrary, defined, singular form, in other words their type. In these conditions, by drawing up maps of peculiar occurrences, by retracing the routes of penetration and the means of propagation of various modes and institutions, it becomes indeed possible to define civilisations, to find their centres of diffusion and even perhaps their points of origin. Lastly, it becomes possible to identify reference points, limits, frontiers and periods, particularly when archaeology and history are enlisted to provide guidance, help and support.

This general fact derives from the very nature and mode of historical propagation of civilisation. This propagation does not just follow any odd path; its fortunes are explicable. It is possible to discern the lines of least resistance and the levels of authority that have been followed. Upon this, hypotheses can be built up with a certain degree of historical probability. If this is legitimate, it is not because imitation is the rule – as Gabriel Tarde believed – but precisely because the borrowing of a certain object of a certain type is by itself – as Durkheim sensed – a relatively distinctive occurrence which can only be explained by the lesser resistance of the borrower and by the authority of those from whom the borrowing is made. Borrowing presupposes indeed a definite genre of historical connections between societies and social facts. These connections can be singled out, and thus contribute to general history.

This propagation reaches in its turn the very nature of civilisation. Here is how. There are two reasons why several elements of social life, which are not strictly political, moral and national, should be limited in this way to a certain number of peoples, linked in their history and their spatial distribution; why civilisations should have frontiers, very much like nations; why they should have a certain permanence in time, a birth, a life and a death, just like the nations they include.

These limits correspond to an inherent quality common to all social phenomena. This quality is manifest even in those phenomena which are not characteristic of a single society, but are nonetheless typical of a certain number of societies, for a shorter or longer period of time. *All social phenomena* have in effect one essential attribute: be they a symbol, a word, an instrument or an institution; be they even best constructed language or science; be they the instrument best adapted to the best and most numerous ends, be they the most rational possible, the most human, *they are still arbitrary*.

All social phenomena are, to some degree, the work of collective will, and who says human will says choice between different possible options. A specific thing, a word, a tale, a form of land management, the internal or external structure of a house, a pottery, a tool, everything has a type, a mode and even, in many cases, in addition to its nature and its model form, its own specific mode of utilisation. The domain of the social is the domain of the modality.

Even the movements themselves, the knot of the tie, the collar and the carriage of the neck which ensues; the gait and the parts of the body which require the wearing of shoes which they then carry – to mention only things which are familiar to us – everything has a form which is at once common to large numbers of people and chosen by them from among other possible forms. This form is, furthermore, one which is found only here or there, and only at such-or-such time periods. Fashion, when considered through time, is quite simply a system of such modalities. Henri Hubert has written some wise words on the 'aspect of a civilisation', on the 'long fields' of the Gauls which still exist today, on the successive forms of roofs, which are not simply determined by geographical causes, as some people would have it. And from all of Gabriel Tarde's writings, I would readily keep as established his subtle moralist comments on 'philoneism' and 'misoneism'.

From this property of collective representations and practices, it follows that the areas of their extension is necessarily finite and relatively fixed – at least until the day humankind will form a single society. For both these chosen collective representations and practices, and the products which embody them, can travel only up to where it is possible and desirable to carry them, and only up to where it is possible and desirable to borrow them (always leaving out the question of periodisation). This arbitrary quality is obviously common only to societies originating from the same stock or the same family of languages, or linked by long-standing contacts, friendly or not (for war, out of necessity, is a great borrower); in other words, societies which have something in common with each other. The limit of an area of civilisation is therefore to be found at the point where come to an end these constant borrowings and evolutions, more or less simultaneous or spontaneous, but always parallel and occurring without too great a separation from a common stock. For example, it is perhaps still possible to talk about a Latin civilisation ... with its Italian or French variants.

This limit, this abrupt halt of an area of civilisation, is very often as sudden as the border of an established society and even of what we call a state. One of the grave lacunae in our studies of collective history, ethnological or other, is that it is much too inclined to observe only coincidences. This gives the impression that only positive phenomena have occurred throughout history. But we need as well to observe non-borrowing, and the rejection of a borrowing, however useful. This research is actually just as fascinating as research into borrowing. For in several cases it is this research that will account for the limits of civilisations, just like the limits of societies. Israel abominated Moab who cooked the lamb in its mother's milk, and that is why people here still abstain from meat on Fridays. The Tuareg only feed on camels' milk and find cows' milk repugnant, just as we abhor mares' milk. The Arctic Indians never knew or wanted to produce for themselves those admirable Eskimo boats, kayaks or umiaks. Conversely the Eskimos have only exceptionally consented to borrow Indian snow shoes. Likewise I have never myself learned to ski, which my young compatriots in the Vosges now regularly do. I have seen movements frozen by the instrument or by habit which have prevented us from using English and German spades with handles, and conversely prevented the English from utilising our own long-handled shovels. One must read in the Ssu-Ma-Ch'en on the debates in the Chinese Court over the Huns' art of horse mounting, and how it was finally permitted, etc.

We can thus see how civilisations are circumscribed, by their capacity for borrowing and expansion, but also by the resistance of the societies which compose them.

* * *

This is how a sociologist, drawing primarily on already existing historical and prehistoric studies and on historical comparisons of civilisations, conceives the history of civilisation in general and that of ethnographic peoples in particular.

In our case, this conception does not date from those unfair and absurd attacks by ethnologists. I will only mention the man who was my brother at work. Henri Hubert prepared a 'Prehistoric Ethnography of Europe', a topic on which he was always an expert. In his book on the Celts that we shall be publishing (in the collection *Evolution de l'humanité*) Hubert identifies their civilisation with that of La Tène. One should go and see his masterpiece, the 'Salle de Mars', which will soon open at the Musée de Saint-Germain. There one will find the history of the whole Neolithic period and of the beginning of metals – a history that is at once chronological, logical and geographical. There one will find an attempt to reach a single solution for these three problems, raised together and simultaneously as they must be.

III. Ordinary meaning of the word civilisation

Following this technical, didactic exposé we are able to return to the ordinary meanings given to the word *civilisation*. In a great many cases it is legitimate to slightly extend its meaning without great scientific error. One can rightly say 'French civilisation', meaning thereby something more than 'French mentality', because this something extends in fact beyond the boundaries of France and even beyond the linguistic limits of French, for example into Flanders or German-speaking Luxembourg. Likewise German culture was until recently still dominant in the Baltic states. Hellenic civilisation, Hellenistic civilisation (we do not understand why its greatness is not appreciated), Byzantine civilisation (about which the same comment applies) all conveyed many objects and ideas over long distances, and encompassed many peoples other than Greeks.

Again, it is also possible to talk about civilisations when large bodies of people have succeeded in creating mentalities, customs, arts and sciences, which spread themselves quite well into the population, which itself forms a state, be it unique or composite. The Oriental Empire, for example, was the seat of the 'Byzantine civilisation'. Marcel Granet rightly talks of a 'Chinese civilisation' inside the frontiers of China. It is equally right to describe as Chinese certain facts beyond these frontiers: everywhere, in fact, reached by Chinese writing, the prestige of the classics and of Chinese drama and music, the symbols of art, the courtesy and art of living that the Chinese possessed, before Europe itself had become polished and orderly. Annam, Korea, Manchuria and Japan are to a greater or lesser degree lands of Chinese civilisation. India for its part is composed of two units and no more. 'India is the Brahman', said Sir Alfred Lyall, and Indian civilisation still persists; through Buddhism it has influenced perhaps the whole of the ancient far-eastern world; the Sanskrit word *nâraka*, 'hell', is used thousands of miles from India, in Indonesia, even in Papua New Guinea. Moreover India and Buddhism are again exerting their influence on us.

One example can give an idea of the complexity of the concrete problem – a problem which a simplistic, naively political and unconsciously abstract and nationalistic history is not even able to raise. The famous frieze and enormous sculptures of the Bayon at Angkor are well known, with their thousands of personages, animals and objects, their four storeys, their ornaments, their celestial, symbolic, earthly and marine personages. But what to make of these great scenes in motion? The whole thing has an indisputable Indo-Khmer appearance. Already a cross-fertilisation, as magnificent as it is curious! But there is more to it: one of the friezes is Buddhist, another represents the Hindu epic; not the Vedic, but rather that of Vishnuism and Sivaism. The interpretation of these last two, as proposed by our French scholars, is beginning to be well accepted. It is the widest of the friezes which presents a hitherto insoluble difficulty. An immense army of thousands of soldiers files past. The priests, chiefs and princes are Hindu or are represented in Indian style. It is thought, but without certainty, to represent the war of Râmayana. The subalterns, the troops, part of the equipment, the weapons, the march, the clothes, the headgear, the movements all come for their part from a distinct, unknown civilisation. The figures (and there is no reason to believe that they are poorly depicted; even stylised, they bear the mark of art and truth) represent a race which bears little resemblance not only to present races, but even to any known pure race. A final series represents daily life and crafts. Some of them already have an Indo-Chinese aspect about them. Thus, Indochina was already from the end of the first millennium AD a 'witch's cauldron', a melting pot for races and civilisations.

From this example emerges a third meaning of the word civilisation: that applied specifically to moral and religious facts. In this sense we are entitled to talk about 'Buddhist civilisation', or more accurately of civilising Buddhism; when it is recalled how Buddhism orchestrates a whole part of the moral and aesthetic life of Indochina, China and Japan, and Korea, and almost all the life, including the political life, of the Tibetans and the Buriats. We are justified to use the expression 'Islamic civilisation', so well can Islam incorporate all into the faithful, from the mere gesture to the innermost being. Even around the idea of the caliphate, Islam has almost formed a political state, and retains so many of its features. Likewise we can correctly speculate about the 'Catholic civilisation' – which is, for itself, 'universal' – of the Occident in medieval times, even when Latin was only the language of the Church and of University. In fact, it is historically more accurate, from the point of view of the contemporaries of this civilisation, to call it Catholic rather than European, given that the notion of Europe did not exist at that time.

Lastly there remains a group of three meanings that are given to the term civilisation, sometimes scientifically and almost always in popular parlance.

By civilisation, philosophers and the public understand 'culture', *Kultur*, the means of rising in the world, of reaching a higher standard of wealth and comfort, of strength and skill, of becoming a civic, civil being, of establishing order and organisation, of imposing civility and politeness, of being distinguished, of savouring and fostering the arts.

Linguists set out from a rather similar idea when they utilise the word 'civilisation' in a double sense. On one hand, they see in the 'languages of civilisation' – Latin, English, German, etc., and now Czech, Serbian, etc. – means of education and transmission of traditions, techniques and sciences, as well as of literary dissemination from quite vast and ancient sources. On the other hand, they contrast these 'civilised' languages with the patois and dialects, the minor languages of small groups and sub-groups, of little-civilised nations, of rural languages par excellence, that is to say languages which are not widespread and hence not refined (this is a probable but unproven inference). For the linguists, the criterion of value and of the expansive character, the common strength and the capacity for transmission of language blend with the quality of the transmitted notions and the transmitted language. Their double definition is not very far removed from our own.

Lastly, there are statesmen, philosophers, the public, and the publicists even more so, who talk about 'the Civilisation'. In nationalist periods, *the Civilisation* is always *their* culture, that of their own nation, for they are generally ignorant of other people's civilisation. In rationalist and generally universalist and cosmopolitan periods, and in the manner of the great religions, *the Civilisation* constitutes a sort of state of things at once ideal and real, both rational and natural, causal and final at the same time, gradually emergent through progress which is never really doubted.

Ultimately, all these meanings of civilisation correspond to an ideal state people have been dreaming about for the past century and a half that they have been thinking politically. This perfect essence has never existed except as a myth, a collective representation. This simultaneously universalist and nationalist belief is actually a distinctive feature of our international and national civilisations of the European West and of non-Indian America. Some would see *the Civilisation* as a perfect nation, corresponding to the 'closed state' of Fichte, autonomous and self-sufficient, whose civilisation and language of civilisation would extend to its political frontiers. Some nations have realised this ideal and others, such as the United States of America, consciously pursue it. Other writers and orators think of *the* human civilisation in the abstract, in the future. Humanity 'in progress' is a commonplace of both philosophy and politics. Lastly, there are others who reconcile the two ideas: the national classes, the nations, the civilisations would only have historical missions vis-à-vis *the Civilisation.* Naturally, this civilisation is always the Occidental one. It is elevated to be a common ideal and at the same time a rational fund of human progress; and, with optimism aiding, it is made the condition of happiness. The nineteenth century mixed the two ideas, and took 'its' civilisation for 'the' civilisation. Every nation and every class has done the same thing, and this has provided material for innumerable pleas.

Nevertheless, it is permissible to believe that the novelty in our life has created something new in this order of things. It seems to us that, in our own era, this time, it is in the facts and no longer in ideology that something such as 'the Civilisation' is being achieved. To begin with, and without nations disappearing or even without all of them being formed, a growing capital of international realities and international ideas is being constituted. The international nature of the facts of civilisation is becoming more intensive. The number of phenomena of this type is increasing; they are spreading, multiplying each other. Their quality is improving. The instrument, such as the above-mentioned shovel, the costume, things of varying degrees of complexity, may possibly remain here and there specific and irrational, picturesque evidence of bygone nations and civilisations. However the machine, the chemical process, cannot. Science dominates everything and, as Leibniz predicted, its language is necessarily human. A new form of communication, of tradition, of description, of the recording of things, even things of sentiment and custom, is becoming universal: this is the cinema. A new form to perpetuate sounds: the gramophone, and a new means for spreading them: the radio-telephone which in less than ten years has emitted all kinds of music, accents, words, information, overcoming all barriers. And we are only at the beginning.

We do not know whether developments will not transform a certain number of elements of civilisation – we have seen this in relation to chemistry and to aviation – into elements of national violence or, what is possibly worse, national pride. Perhaps nations will again have no scruples in breaking away from the humanity which nourishes them and by which they are increasingly elevated. But it is certain that hitherto unheard-of permutations are being established; it is certain that, with nations and civilisations continuing to exist, the number of features they share in common will increase and the forms of each feature will resemble the others more and more closely. This is because the common base they all share is every day growing in number, weight and quality, and reaching further afield at an accelerated pace. What is more, some elements of this new civilisation actually originate from populations which until only recently or still today are excluded or severed from it; the success of the primitive arts, including music, demonstrates that the history of all this will follow many unknown paths.

Let us end with this notion of *common fund*, of *overall achievement of societies and civilisations*. This, in my opinion, is what the notion of 'the Civilisation' corresponds to, as the extent of fusion and not as the principle of civilisations. These civilisations are nothing if they are not cherished and developed by the nations which bear them. But just as within nations, where science, industries, the arts, 'distinction' itself are ceasing to be the heritage of a small class of people to become, in the great nations, a sort of common privilege, likewise the best features of these civilisations will become the common property of more and more numerous social groups. The poet or the historian may mourn the loss of local flavours; ways to save them will possibly be found. But the capital of humankind will grow in all cases. The products, the organisation of the land and the coasts are all increasingly rationally established, exploited for a market which is now global. One is entitled to say that *this* is where *the civilisation* resides. Indisputably, all the nations and civilisations at present are tending towards *more*; *more powerful*, *more general* and *more rational* (the last two terms are reciprocal in so far as, symbolism excepted, humans commune only in the rational and the real).

And this *more* is obviously increasingly widespread, better understood and above all definitively retained by more and more people. Charles Seignobos used to say that a civilisation is roads, harbours and quays. In this sally, he was isolating capital from the industry that creates it. Included in this must also be the capital of reason which created it: 'pure reason', 'practical reason', 'power of judgement', to use Kant's language. This notion of a growing achievement, of intellectual and material wealth shared by an increasingly reasonable humankind, is, we sincerely believe, factually based. It makes it possible to assess civilisations sociologically, to appreciate the contributions of a given nation to civilisation, without it being necessary to make value judgements, be it on nations, on civilisations, or on *the Civilisation*. For *the Civilisation*, any more than progress, does not necessarily lead to good or happiness.

But I leave to Alfredo Niceforo the task of discussing the question of value judgements on these matters.

5. 'Camped on marsupial skins around a fire; weapons ready for attack'. P.105.ACH1. New South Wales, Australia. Mounted Haddon Collection (CUMAA©). 'Techniques of sleep. The notion that going to sleep is something natural is totally inaccurate' (Mauss, text 9).

Text 8

Fragment of a Plan of General Descriptive Sociology (on Tradition) (1934, extract)

M. Mauss, 1934, 'Fragment d'un plan de sociologie générale descriptive. Classification et méthode d'observation des phénomènes généraux de la vie sociale dans les sociétés de type archaïque (phénomènes généraux spécifiques de la vie intérieure de la société)', *Annales sociologiques* série A, 1:1–56. (Extract pp. 33–34.)

Translated by Nathan Schlanger

As its title indicates, this essay represents yet another attempt by Mauss to systematise the sociological domain, very much in Durkheim's spirit. Phenomena of 'general sociology' (which are distinguished from and follow the 'special sociologies' – religion, morals, law, economics, techniques) express the solidarity of the latter, and show that 'societies form systems'. At the intra-social level, Mauss focused on the key phenomenon of social cohesion and discussed its possible variants, the discipline and authority that establish it, and also the transmission of social cohesion through education and tradition. In the brief extract translated here, Mauss touches suggestively on the 'traditional' nature of technical actions, and also on the linguistic, symbolic and material dimensions of what he calls 'significant movements'.

It is possible to distinguish between two kinds of traditions. Firstly there is oral tradition, which seems to be the only kind in our societies, and which has certainly been a characteristic feature of humanity since its origins. I shall not dwell any further on this self-evident topic. Besides this oral tradition, there is another one, more basic perhaps, which is generally mistaken for imitation. The use of oral symbols is but one case of the use of symbols: any traditional practice, endowed with a form and transmitted through that form, can in some measure be regarded as symbolic. When one generation hands down to the next the technical knowledge of its manual and bodily actions, as much authority and social tradition is involved as when transmission occurs through language. In this there is truly tradition, and continuity. The important deed is the handing over of science, knowledge and power from master to pupil; everything can perpetuate itself in this way.

These are rather the intellectual forms of thought that depend on language in order to be communicated. The other forms of moral and material life are transmitted through direct contact, and this communication necessarily takes place in a context of authority. This applies equally as far as the forms of emotional life are concerned. Moral and religious feelings, as well as technical or aesthetic activities, and so on, are imposed by elders on youth, by leaders on followers, by one individual on another. What is labelled imitation in the realm of individual psychology, and what is worthy of the name tradition in social psychology, boils down to this way in which things become established in ancient societies. Wisdom, etiquette, ability, skill – even in sports – ultimately find expression in two ways: these are proverbs, opinions and saws, dictamina, precepts, myths, tales, riddles, etc., on the one hand; on the other, significant movements, and finally a series of movements the success of which is believed or known to be certain precisely because they form a chain where the first is a sign for the others to follow. Since their value as signs is known not only by the agent but equally by all those present, and since they are construed as causes by the agent and by the spectators at the same time, they therefore become symbolic gestures which are simultaneously real, physically effective gestures. Furthermore, the fact that this physical effectiveness is confounded with religious and moral effectiveness leads us to conceive that in such societies, the symbols of procedure and ritual are of the same kind as those of meals, walking and posture, etc. I shall come back to this point.

Thus tradition extends to everything, and it is at least very powerful. Its omnipotence, the power of its constraining character, in Durkheim's thinking, has been contested (Moszkowski, the Revd Schmidt, etc.). Abstract discussion here is useless. Observation and a sense of relative importance is what we need.

Text 9

Techniques of the Body (1935)

M. Mauss, 1935, 'Les techniques du corps', *Journal de psychologie* 32:271–93. Translated by Ben Brewster and previously published in *Economy and Society* (1973) 2/1:70–88.

All notes and references follow Brewster.

Mauss's presidential address to the Société de Psychologie Française in May 1934, subsequently published in the society's Journal, was part of the ongoing interdisciplinary contacts he had initiated with psychology since the early 1920s, seeking notably to move beyond the residual antagonisms left by Durkheim. In this context, and through dialogues with Meyerson, Georges Dumas and other scholars, Mauss elaborated his famous conception of 'totality', as generally applied to the *homme total* and more specifically to the expression of sentiments, the idea of death and indeed the gift. In this text, Mauss revealingly disclosed the mental processes by which he came to reach this notion of 'techniques of the body', and how he rescued the phenomena thus designated from the unsatisfactory category of 'divers' by realising that there exists a range of techniques without instruments. Mauss went on to propose that in their corporeal and technical *habitus*, individuals are 'total' human beings, setting in motion the biological, psychological and sociological dimensions of their being.

Chapter One: The Notion of Techniques of the Body

I deliberately say techniques of the body in the plural because it is possible to produce a theory of *the* technique of the body in the singular on the basis of a study, an exposition, a description pure and simple of techniques of the body in the plural. By this expression I mean the ways in which from society to society men know how to use their bodies. In any case, it is essential to move from the concrete to the abstract and not the other way round.

I want to convey to you what I believe is one of the parts of my teaching which is not to be found elsewhere, that I have rehearsed in a course of lectures on descriptive ethnology (the books containing the *Summary Instructions* and *Instructions for Ethnographers* are to be published) and have tried out several times in my teaching at the Institut d'Ethnologie of the University of Paris.

When a natural science makes advances, it only ever does so in the direction of the concrete, and always in the direction of the unknown. Now the unknown is found at the frontiers of the sciences, where the professors are at each other's throats, as Goethe put it (though Goethe was not so polite). It is generally in these ill-demarcated domains that the urgent problems lie. Moreover, these uncleared lands are marked. In the natural sciences at present, there is always one obnoxious rubric. There is always a moment when, the science of certain facts not being yet reduced into concepts, the facts not even being organically grouped together, these masses of facts receive that posting of ignorance: 'Miscellaneous'. This is where we have to penetrate. We can be certain that this is where there are truths to be discovered: first, because we know that we are ignorant, and second, because we have a lively sense of the quantity of the facts. For many years in my course in descriptive ethnology, I have had to teach in the shadow of the disgrace and opprobrium of the 'miscellaneous' in a matter in which in ethnography this rubric 'miscellaneous' was truly heteroclite. I was well aware that walking or swimming, for example, and all sorts of things of the same type, are specific to determinate societies; that the Polynesians do not swim as we do, that my generation did not swim as the present generation does. But what social phenomena did these represent? They were 'miscellaneous' social phenomena, and, as this rubric is a horror, I have often thought about this 'miscellaneous', at least as often as I have been obliged to discuss it and often in between times.

Forgive me if, in order to give this notion of techniques of the body shape for you, I tell you about the occasions on which I pursued this general problem and how I managed to pose it clearly. It was a series of steps consciously and unconsciously taken.

First, in 1898, I came into contact with someone whose initials I still know, but whose name I can no longer remember.[1] I have been too lazy to look it up. It was the man who wrote an excellent article on 'Swimming' for the 1902 edition of the *Encyclopaedia Britannica*, then in preparation. (The articles on 'Swimming' in the two later editions are not so good.) He revealed to me the historical and ethnographical interest of the question. It was a startingpoint, an observational framework. Subsequently – I noticed it myself – we have seen swimming techniques undergo a change, in our generation's lifetime. An example will put us in the picture straight away: us, the psychologists, as well as the biologists and sociologists. Previously we were taught to dive after having learnt to swim. And when we were learning to dive, we were taught to close our eyes and then to open them under water. Today the technique is the other way round. The whole training begins by getting the children used to keeping their eyes open under water. Thus, even before they can swim, particular care is taken to get the children to control their dangerous but instinctive ocular reflexes, before all else they are familiarised with the water, their fears are suppressed, a certain confidence is created, suspensions and movements are selected. Hence there is a technique of diving and a technique of education in diving which have been discovered in my day. And you can see that it really is a technical education and, as in every technique, there is an apprenticeship in swimming. On the other hand, here our generation has witnessed a complete change in technique: we have seen the breaststroke with the head out of the water replaced by the different sorts of crawl. Moreover, the habit of swallowing water and spitting it out again has gone. In my day swimmers thought of themselves as a kind of steamboat. It was stupid, but in fact I still do this: I cannot get rid of my technique. Here then we have a specific technique of the body, a gymnic art perfected in our own day.

But this specificity is characteristic of all techniques. An example: during the War I was able to make many observations on this specificity of techniques. E.g. the technique of *digging*. The English troops I was with did not know how to use French spades, which forced us to change 8,000 spades a division when we relieved a French division, and vice versa. This plainly shows that a manual knack can only be learnt slowly. Every technique properly so-called has its own form.

But the same is true of every attitude of the body. Each society has its own special habits. In the same period I had many opportunities to note the differences between the various armies. An anecdote about *marching*. You all know that the British infantry marches with a different step from our own: with a different frequency and a different stride. For the moment I am not talking about the English swing or the action of the knees, etc. The Worcester Regiment, having achieved considerable glory alongside French infantry in the Battle of the Aisne, requested Royal permission to have French trumpets and drums, a band of French buglers and drummers. The result was not very encouraging. For nearly six months, in the streets of Bailleul, long after the Battle of the Aisne, I often saw the following sight: the regiment had preserved its English march but had set it to a French rhythm. It even had at the head of its band a little French light infantry regimental sergeant major who could blow the bugle and sound the march even better than his men. The unfortunate regiment of tall Englishmen could not march. Their gait was completely at odds. When they tried to march in step, the music would be out of step. With the result that the Worcester Regiment was forced to give up its French buglers. In fact, the buglecalls adopted army by army earlier, in the Crimean War, were the calls 'at ease', 'retreat', etc. Thus I saw in a very precise and frequent fashion, not only with the ordinary march, but also at the double and so on, the differences in elementary as well as sporting techniques between the English and the French. Prince Curt Sachs, who is living here in France at present, made the same observation. He has discussed it in several of his lectures. He could recognise the gait of an Englishman and a Frenchman from a long distance.

But these were only approaches to the subject. A kind of revelation came to me in hospital. I was ill in New York. I wondered where previously I had seen girls walking as my nurses walked. I had the time to think about it. At last I realised that it was at the cinema. Returning to France, I noticed how common this gait was, especially in Paris; the girls were French and they too were walking in this way. In fact, American walking fashions had begun to arrive over here, thanks to the cinema. This was an idea I could generalise. The positions of the arms and hands while walking form a social idiosyncrasy, they are not simply a product of some purely individual, almost completely psychical arrangements and mechanisms. For example: I think I can also recognise a girl who has been raised in a convent. In general she will walk with her fists closed. And I can still remember my third-form teacher shouting at me: 'Idiot! why do you walk around the whole time with your hands flapping wide open?' Thus there exists an education in walking, too.

Another example: there are polite and impolite *positions for the hands* at rest. Thus you can be certain that if a child at table keeps his elbows in when he is not eating he is English. A young Frenchman has no idea how to sit up straight; his elbows stick out sideways; he puts them on the table, and so on.

Finally, in *running* too, I have seen, you all have seen, the change in technique. Imagine that my gymnastics teacher, one of the top graduates of Joinville around 1860, taught me to run with my fists close to my chest: a movement completely contradictory to all running movements; I had to see the professional runners of 1890 before I realised the necessity of running in a different fashion.

Hence I have had this notion of the social nature of the *'habitus'* for many years. Please note that I use the Latin word – it should be understood in France – habitus. The word translates infinitely better than *'habitude'* (habit or custom), the 'exis', the 'acquired ability' and 'faculty' of Aristotle (who was a psychologist). It does not designate those metaphysical *habitudes*, that mysterious 'memory', the subjects of volumes or short and famous theses. These 'habits' do not just vary with individuals and their imitations, they vary especially between societies, educations, proprieties and fashions, prestige. In them we should see the techniques and work of collective and individual practical reason rather than, in the ordinary way, merely the soul and its repetitive faculties.

Thus everything moved me towards the position that we in this Society are among those who have adopted, following Comte's example: the position of Georges Dumas, for example, who, in the constant relations between the biological and the sociological, leaves but little room for the psychological mediator. And I concluded that it was not possible to have a clear idea of all these facts about running, swimming, etc., unless one introduced a triple consideration instead of a single consideration, be it mechanical and physical, like an anatomical and physiological theory of walking, or on the contrary psychological or sociological. It is the triple viewpoint, that of the 'total man', that is needed.

Lastly, another series of facts impressed itself upon me. In all these elements of the art of using the human body, the facts of *education* were dominant. The notion of education could be superimposed on that of imitation. For there are particular children with very strong imitative faculties, others with very weak ones, but all of them go through the same education, such that we can understand the continuity of the concatenations. What takes place is a prestigious imitation. The child, the adult, imitates actions which have succeeded and which he has seen successfully performed by people in whom he has confidence and who have authority over him. The action is imposed from without, from above, even if it is an exclusively biological action, involving his body. The individual borrows the series of movements which constitute it from the action executed in front of him or with him by others.

It is precisely this notion of the prestige of the person who performs the ordered, authorised, tested action vis-à-vis the imitating individual that contains all the social element. The imitative action which follows contains the psychological element and the biological element. But the whole, the ensemble, is conditioned by the three elements indissolubly mixed together.

All this is easily linked to a number of other facts. In a book by Elsdon Best that reached here in 1925 there is a remarkable document on the way Maori women (New Zealand) walk. (Do not say that they are primitives, for in some ways I think they are superior to the Celts and Germans.) 'Native women adopted a peculiar gait [the English word is delightful] that was acquired in youth, a loosejointed swinging of the hips that looks ungainly to us, but was admired by the Maori. Mothers drilled their daughters in this accomplishment, termed *onioni*, and I have heard a mother say to her girl: "*Ha! Kaore koe e onioni*" [you are not doing the *onioni*] when the young one was neglecting to practise the gait'.[2] This was an acquired, not a natural way of walking. To sum up, there is perhaps no 'natural way' for the adult. A fortiori when other technical facts intervene: to take ourselves, the fact that we wear shoes to walk transforms the positions of our feet: we feel it sure enough when we walk without them.

On the other hand, this same basic question arose for me in a different region, vis-à-vis all the notions concerning magical power, beliefs in the not only physical but also moral, magical and ritual effectiveness of certain actions. Here I am perhaps even more on my own terrain than on the adventurous terrain of the psychophysiology of modes of walking, which is a risky one for me in this company.

Here is a more 'primitive' fact, Australian this time: a ritual formula both for hunting and for running. As you will know, the Australian manages to outrun kangaroos, emus, and wild dogs. He manages to catch the possum or phalanger at the top of its tree, even though the animal puts up a remarkable resistance.

One of these running rituals, observed a hundred years ago, is that of the hunt for the dingo or wild dog among the tribes near Adelaide. The hunter constantly shouts the following formula:

Strike (him, i.e. the dingo) with the tuft of eagle feathers (used in initiation, etc.)
Strike (him) with the girdle
Strike (him) with the string round the head
Strike (him) with the blood of circumcision
Strike (him) with the blood of the arm
Strike (him) with menstrual blood
Send (him) to sleep, etc.[3]

In another ceremony, that of the possum hunt, the individual carries in his mouth a piece of rock crystal *(kawemukka)*, a particularly magical stone, and chants a formula of the same kind, and it is with this support that he is able to dislodge the possum, that he climbs the tree and can stay hanging on to it by his belt, that he can outlast and catch and kill this difficult prey.

The relations between magical procedures and hunting techniques are clear, too universal to need stressing.

The psychological phenomenon I am reporting at this moment is clearly only too easy to know and understand from the normal point of view of the sociologist. But what I want to get at now is the confidence, the psychological *momentum* that can be linked to an action which is primarily a fact of biological resistance, obtained thanks to some words and a magical object.

Technical actions, physical actions, magicoreligious actions are confused for the actor. These are the elements I had at my disposal.

* * *

All this did not satisfy me. I saw how everything could be described, but not how it could be organised; I did not know what name, what title to give it all. It was very simple, I just had to refer to the division of traditional actions into techniques and rites, which I believe to be well founded. All these modes of action were techniques, the techniques of the body.

I made, and went on making for several years, the fundamental mistake of thinking that there is technique only when there is an instrument. I had to go back to ancient notions, to the Platonic position on technique, for Plato spoke of a technique of music and in particular of a technique of the dance, and extend these notions.

I call technique an action which is *effective* and *traditional* (and you will see that in this it is no different from a magical, religious or symbolic action). It has to be *effective* and *traditional*. There is no technique and no transmission in the absence of tradition. This above all is what distinguishes humans from the animals: the transmission of their techniques and very probably their oral transmission.

it is possible not only to classify techniques, but also to classify their variations by age and sex.

Having established this classification, which cuts across all classes of society, we can now glimpse a third one.

3. *Classification of techniques of the body according to efficiency*

The techniques of the body can be classified according to their efficiency, i.e. according to the results of training. Training, like the assembly of a machine, is the search for, the acquisition of efficiency. Here it is a human efficiency. These techniques are thus human norms of human training. These procedures that we apply to animals men voluntarily apply to themselves and to their children. The latter are probably the first beings to have been trained in this way, before all the animals, which first had to be tamed. As a result I could to a certain extent compare these techniques, them and their transmission, to training systems, and rank them in the order of their effectiveness.

This is the place for the notion of dexterity, so important in psychology, as well as in sociology. But in French we only have the poor term *'habile'* which is a bad translation of the Latin word *'habilis'* far better designating those people with a sense of the adaptation of all their well-coordinated movements to a goal, who are practised, who 'know what they are up to'. The English notions of 'craft' or 'cleverness' (skill, presence of mind and habit combined) imply competence at something. Once again we are clearly in the domain of techniques.

4. *Transmission of the form of the techniques*

One last viewpoint: the teaching of techniques being essential, we can classify them according to the nature of this education and training. Here is a new field of studies: masses of details which have not been observed, but should be, constitute the physical education of all ages and both sexes. The child's education is full of so-called details, which are really essential. Take the problem of ambidextrousness for example: our observations of the movements of the right hand and of the left hand are poor and we do not know how much all of them are acquired. A pious Muslim can easily be recognised: even when he has a knife and fork (which is rarely), he will go to any lengths to avoid using anything but his right hand. He must never touch his food with his left hand, or certain parts of his body with his right. To know why he does not make a certain gesture and does make a certain other gesture neither the physiology nor the psychology of motor asymmetry in man is enough, it is also necessary to know the traditions which impose it. Robert Hertz has posed this problem correctly.[8] But reflections of this and other kinds can be applied whenever there is a social choice of the principles of movements.

There are grounds for studying all the modes of training, imitation and especially those fundamental fashions that can be called the modes of life, the *modes*, the *tonus* the 'matter', the 'manners', the 'way'.

Here is the first classification, or rather, four viewpoints.

Chapter Three: A Biographical List of the Techniques of the Body

Another quite different classification is, I would not say more logical, but easier for the observer. It is a simple list. I had thought of presenting to you a series of small tables, of the kind American professors construct. I shall simply follow more or less the ages of man, the normal biography of an individual, as an arrangement of the techniques of the body which concern him or which he is taught.

1. *Techniques of birth and obstetrics*

The facts are rather little known, and much of the classical information is disputable.[5] Among the best is that of Walter Roth on the Australian tribes of Queensland and on those of British Guiana.[6]

The forms of obstetrics are very variable. The infant Buddha was born with his mother Mâya upright and clinging to the branch of a tree. She gave birth standing up. Indian women still in the main give birth in this position. Something we think of as normal, like giving birth lying on one's back, is no more normal than doing so in other positions, e.g. on all fours. There are techniques of giving birth, both on the mother's part and on that of her helpers, of holding the baby, cutting and tying the umbilical cord, caring for the mother, caring for the child. Here are quite a number of questions of some importance. And here are some more: the choice of the child, the exposure of weaklings, the killing of twins are decisive moments in the history of a race. In ancient history and in other civilisations, the recognition of the child is a crucial event.

2. *Techniques of infancy*

Rearing and feeding the child. Attitudes of the two interrelated beings: mother and child. Take the child, suckling, etc., carrying, etc. The history of carrying is very important. A child carried next to its mother's skin for two or three years has a quite different attitude to its mother from that of a child not so carried;[7] it has a contact with its mother utterly unlike our children's. It clings to her neck, her shoulder, it sits astride her hip. This remarkable gymnastics is essential throughout its life. And there is another gymnastics for the mother carrying it. It even seems that psychical states arise here which have disappeared from infancy with us. There are sexual contacts, skin contacts, etc.

Weaning. Takes a long time, usually two or three years. The obligation to suckle, sometimes even to suckle animals. It takes a long time for the mother's milk to run dry. Besides this there are relations between weaning and reproduction, suspensions of reproduction during weaning.[8]

Mankind can more or less be divided into people with cradles and people without. For there are techniques of the body which presuppose an instrument. Countries with cradles include almost all the peoples of the two Northern hemispheres, those of the Andean region, and also a certain number of Central African populations. In these last two groups, the use of the cradle coincides with a cranial deformation (which perhaps has serious physiological consequences).

The weaned child. It can eat and drink; it is taught to walk; it is trained in vision, hearing, in a sense of rhythm and form and movement, often for dancing and music. It acquires the notions and practices of physical exercise and breathing. It takes certain postures which are often imposed on it.

3. *Techniques of adolescence*

To be observed with men in particular. Less important with girls in those societies to whose study a course in Ethnology is devoted. The big moment in the education of the body is, in fact, the moment of initiation. Because of the way our boys and girls are brought up we imagine that both acquire the same manners and postures and receive the same training everywhere. The idea is already erroneous about ourselves – and it is totally false in so-called primitive countries. Moreover, we describe the facts as if something like our own school, beginning straight away and intended to protect the child and train it for life, had always and everywhere existed. The opposite is the rule. For example: in all black societies the education of the boy intensifies around the age of puberty, while that of women remains traditional, so to speak. There is no school for women. They are at school with their mothers and are formed there continuously, moving directly, with few exceptions, to the married state. The male child enters the society of men where he learns his profession, especially the profession of arms. However, for men as well as women, the decisive moment is that of adolescence. It is at this moment that they learn definitively the techniques of the body that they will retain for the whole of their adult lives.

4. *Techniques of adult life*

To list these we can run through the various moments of the day among which coordinated movements and suspensions of movement are distributed.

We can distinguish sleep and waking, and in waking, rest and activity.

a. *Techniques of sleep.* The notion that going to sleep is something natural is totally inaccurate. I can tell you that the War taught me to sleep anywhere, on heaps of stones for example, but that I have never been able to change my bed without a moment of insomnia: only on the second night can I go to sleep quickly.

One thing is very simple: it is possible to distinguish between those societies that have nothing to sleep on except the 'floor', and those that have instrumental assistance. The 'civilisation of latitude 15°' discussed by Graebner[9] is characterised among other things by its use of a bench for the neck. This neck-rest is often a totem, sometimes carved with squatting figures of men and totemic animals. There are people with mats and people without (Asia, Oceania, part of America). There are people with pillows and people without. There are populations which lie very close together in a ring to sleep, round a fire, or even without a fire. There are primitive ways of getting warm and keeping the feet warm. The Fuegians, who live in a very cold region, cannot warm their feet while they are asleep having only one blanket of skin *(guanaco)*. Finally there is sleep standing up. The Masai can sleep on their feet. I have slept standing up in the mountains. I have often slept on a horse, even sometimes a moving horse: the horse was more intelligent than I was. The old chroniclers of the invasions picture the Huns and Mongols sleeping on horseback. This is still true, and their riders' sleeping does not stop the horses' progress.

There is the use of coverings. People who sleep covered and uncovered. There is the hammock and the way of sleeping hanging up.

Here are a large number of practices which are both techniques of the body and also have profound biological echoes and effects. All this can and must be observed on the ground; hundreds of things still remain to be discovered.

b. *Techniques of rest.* Rest can be perfect rest or a mere suspension of activity: lying down, sitting, squatting, etc. Try squatting. You will realise the torture that a Moroccan meal, for example, eaten according to all the rituals, would cause you. The way of sitting down is fundamental. You can distinguish squatting mankind and sitting mankind. And, in the latter, people with benches and people without benches and daises; people with chairs and people without chairs. Wooden chairs supported by crouching figures are widespread, curiously enough, in all the regions at fifteen degrees of latitude North and along the Equator in both continents.[10] There are people who have tables and people who do not. The table, the Greek *trapeza*, is far from universal. Normally it is still a carpet, a mat, throughout the East. This is all complicated, for these forms of rest include meals, conversation, etc. Certain societies take their rest in very peculiar positions. Thus, the whole of Nilotic Africa and part of the Chad region, all the way to Tanganyika, is populated by men who rest in the fields like storks. Some manage to rest on one foot without a pole, others lean on a stick. These resting techniques form real characteristics of civilisations, common to a large number of them, to whole families of peoples. Nothing seems more natural to the psychologists; I do not know if they would quite agree with me, but I believe that these postures in the savannah are due to the height of the grasses there and the functions of shepherd or sentry, etc.; they are laboriously acquired by education and preserved.

You have active, generally aesthetic rest; thus even dancing at rest is frequent, etc. I shall return to this.

c. *Techniques of activity, of movement.* By definition, rest is the absence of movements, movement the absence of rest. Here is a straightforward list: movements of the whole body: climbing; trampling; walking.

Walking. The *habitus* of the body being upright while walking, breathing, rhythm of the walk, swinging the fists, the elbows, progression with the trunk in advance of the body or by advancing either side of the body alternately (we have got accustomed to moving all the body forward at once). Feet in or out. Extension of the leg. We laugh at the 'goose step'. It is the way the German Army can obtain the maximum extension of the leg, given in particular that all Northerners, high on their legs, like to make steps as long as possible. In the absence of these exercises, we Frenchmen remain more or less knockkneed. Here is one of those idiosyncrasies which are simultaneously matters of race, of individual mentality and of collective mentality. Techniques such as those of the about-turn are among the most curious. The about-turn 'on principle' English style is so different from our own that it takes considerable study to master it.

Running. Position of the feet, position of the arms, breathing, running magic, endurance. In Washington I saw the chief of the Fire Fraternity of the Hopi Indians who had arrived with four of his men to protest against the prohibition of the use of certain alcoholic liquors in their ceremonies. He was certainly the best runner in the world. He had run 250 miles without stopping. All the Pueblos are accustomed to prodigious physical feats of all kinds. [Henri] Hubert, who had seen them, compared them physically with Japanese athletes. This same Indian was an incomparable dancer.

Finally we reach techniques of active rest which are not simply a matter of aesthetics, but also of bodily games.

Dancing. You have perhaps attended the lectures of Erich Maria von Hornbostel and M. Curt Sachs. I recommend to you the latter's very fine history of dancing.[11] I accept their division into dances at rest and dances in action.[12] I am less prepared to accept their hypothesis about the distribution of these dances. They are victims to the fundamental error which is the mainstay of a whole section of sociology. There are supposed to be societies with exclusively masculine descent and others with exclusively uterine descent. The uterine ones, being feminised, tend to dance on the spot; the others, with descent by the male, take their pleasure in moving about.

Curt Sachs has better classified these dances into extravert and introvert dances.[13] We are plunged straight into psychoanalysis, which is probably quite well founded here. In fact the sociologist has to see things in a more complex way. Thus, the Polynesians and in particular the Maori, shake very greatly, even on the spot, or move about very much when they have the space to do so.

Men's dancing and women's dancing should be distinguished, for they are often opposed.

Lastly we should realise that dancing in a partner's arms is a product of modern European civilisation. Which shows you that things we find natural are historical. Moreover, they horrify everyone in the world but ourselves.

I move on to the techniques of the body which are also a function of trades and part of vocations or more complex techniques.

Jumping. We have witnessed a transformation of jumping techniques. We all jumped from a springboard and, once again, fullface. I am glad to say that this has stopped. Now people jump, fortunately, from one side. Jumping lengthways, sideways, up and down. Standing jump, polejump. Here we return to the objects of the reflections of our friends Wolfgang Köhler, Paul Guillaume and Ignace Meyerson: the comparative psychology of humans and animals. I won't say anything more about it. These techniques are infinitely variable.

Climbing. I can tell you that I'm very bad at climbing trees, though reasonable on mountains and rocks. A difference of education and hence of method.

A method of getting up trees with a belt encircling the tree and the body is crucial among all so-called primitives. But we do not have the use of this belt. We see telegraph workers climbing with crampons, but no belt. This procedure should be taught there.[14]

The history of mountaineering methods is very noteworthy. It has made fabulous progress in my lifetime.

Descent. Nothing makes me so dizzy as watching a Kabyle going downstairs in Turkish slippers *(babouches)*. How can he keep his feet without the slippers coming off? I have tried to see, to do it, but I can't understand. Nor can I understand how women can walk in high heels. Thus there is a lot even to be observed, let alone compared.

Swimming. I have told you what I think. Diving, swimming; use of supplementary means; airfloats, planks, etc. We are on the way to the invention of navigation. I was one of those who criticised the de Rougés book on Australia, demonstrated their plagiarisms, believed they were grossly inaccurate. Along with so many others I held their story for a fable: they had seen the Niol-Niol (N.W. Australia) riding cavalcades of great seaturtles. But now we have excellent photographs in which these people can be seen riding turtles. In the same way Robert Sutherland Rattray noted the story of pieces of wood on which people swim among the Ashanti.[15] Moreover, it has been confirmed for the natives of almost all the lagoons of Guinea, Porto-Novo in our own colonies.

Forceful movements. Pushing, pulling, lifting. Everyone knows what a backheave is. It is an acquired technique, not just a series of movements. Throwing, up or along the ground, etc.; the way of holding the object to be thrown between the fingers is noteworthy and undergoes great variation.

Holding. Holding between the teeth. Use of the toes, the armpit, etc.

This study of mechanical movements has got off to a good start. It is the formation of mechanical 'pairs of elements' with the body. You will recall Franz

Reuleaux's great theory about the formation of these pairs of elements.[16] And here the great name of Louis-Hubert Farabeuf will not be forgotten. As soon as I use my fist, and a fortiori, when a man had a 'Chellean hand-axe' in his hand, these 'pairs of elements' are formed.

This is the place for conjuring tricks, sleight of hand, athletics, acrobatics, etc. I must tell you that I had and still have a great admiration for jugglers and gymnasts.

d. *Techniques of care for the body. Rubbing, washing, soaping.* This dossier is hardly a day old. The inventors of soap were not the Ancients, they did not use it. It was the Gauls. And on the other hand, independently, in the whole of Central and North East of South America they soaped themselves with *quillaia* bark or 'brazil', hence the name of the empire.

Care of the mouth. Coughing and spitting technique. Here is a personal observation. A little girl did not know how to spit and this made every cold she had much worse. I made inquiries. In her father's village and in her father's family in particular, in Berry, people do not know how to spit. I taught her to spit. I gave her four sous per spit. As she was saving up for a bicycle she learnt to spit. She is the first person in her family who knows how to spit.

Hygiene in the needs of nature. Here I could list innumerable facts for you.

e. *Consumption techniques. Eating.* You will remember the story Herald Høffding repeats about the Shah of Persia. The Shah was the guest of Napoleon III and insisted on eating with his fingers. The Emperor urged him to use a golden fork. 'You don't know what a pleasure you are missing,' the Shah replied.

Absence and use of knives. An enormous factual error is made by William S. McGee who believed he had observed that the Seri (Indians of the Madeleine Peninsula, California), having no notion of knives, were the most primitive human beings. They did not have knives for eating, that is all.[17]

Drinking. It would be very useful to teach children to drink straight from the source, the fountain, etc., or from puddles of water, etc., to pour their drinks straight down their throats, etc.

f. *Techniques of reproduction.* Nothing is more technical than sexual positions. Very few writers have had the courage to discuss this question. We should be grateful to M. Friedrich Saloman Krauss for having published his great collection of *Anthropophyteia*.[18] Consider for example the technique of the sexual position consisting of this: the woman's legs hang by the knees from the man's elbows. It is a technique specific to the whole Pacific, from Australia to lower Peru, via the Bering Straits – very rare, so to speak, elsewhere.

There are all the techniques of normal and abnormal sexual acts. Contact of the sexual organs, mingling of breath, kisses, etc. Here sexual techniques and sexual morals are closely related.

g. Lastly there are the *techniques of the care of the abnormal:* massages, etc. But let us move on.

Chapter Four: General Considerations

General questions may perhaps be of more interest to you than these lists of techniques that I have paraded before you at rather too great a length.

What emerges very clearly from them is the fact that we are everywhere faced with physiopsychosociological assemblages of series of actions. These actions are more or less habitual and more or less ancient in the life of the individual and the history of the society.

Let us go further: one of the reasons why these series may more easily be assembled in the individual is precisely because they are assembled by and for social authority. As a corporal this is how I taught the reason for exercise in close order, marching four abreast and in step. I ordered the soldiers not to march in step drawn up in ranks and in two files four abreast, and I obliged the squad to pass between two of the trees in the courtyard. They marched on top of one another. They realised that what they were being made to do was not so stupid. In group life as a whole there is a kind of education of movements in close order.

In every society, everyone knows and has to know and learn what he has to do in all conditions. Naturally, social life is not exempt from stupidity and abnormalities. Error may be a principle. The French Navy only recently began to teach its sailors to swim. But example and order, that is the principle. Hence there is a strong sociological causality in all these facts. I hope you will accept that I am right.

On the other hand, since these are movements of the body, this all presupposes an enormous biological and physiological apparatus. What is the breadth of the linking psychological cogwheel? I deliberately say cogwheel. A Comtean would say that there is no gap between the social and the biological. What I can tell you is that here I see psychological facts as connecting cogs and not as causes, except in moments of creation or reform. Cases of invention, of laying down principles, are rare. Cases of adaptation are an individual psychological matter.

But in general they are governed circumstances of life in common, of contact.

On the other hand there are two big questions on the agenda for psychology: the question of individual capacities, of technical orientation, and the question of salient features, of biotypology, which may concur with the brief investigations I have just made. The great advances of psychology in the last few years have not, in my opinion, been made vis-à-vis each of the so-called faculties of psychology, but in psychotechnics, arid in the analysis of psychological 'wholes'.

Here the ethnologist comes up against the big questions of the psychical possibilities of such a race and such a biology of such a people. These are fundamental questions. I believe that here, too, whatever the appearances, we are dealing with biologicosociological phenomena. I think that the basic education in all these techniques consists of an adaptation of the body to their use. For example, the great tests of stoicism, etc., which constitute initiation for the majority of mankind, have as their aim to teach composure, resistance, seriousness, presence of mind, dignity, etc. The main utility I see in my erstwhile mountaineering was this education of my composure, which enabled me to sleep upright on the narrowest ledge overlooking an abyss.

I believe that this whole notion of the education of races that are selected on the basis of a determinate efficiency is one of the fundamental moments of history itself: education of the vision, education in walking–ascending, descending, running. It consists especially of education in composure. And the latter is above all a retarding mechanism, a mechanism inhibiting disorderly movements; this retardation subsequently allows a coordinated response of coordinated movements setting off in the direction of a chosen goal. This resistance to emotional seizure is something fundamental in social and mental life. It separates out, it even classifies the so-called primitive societies; according to whether they display more brutal, unreflected, unconscious reactions or on the contrary more isolated, precise actions governed by a clear consciousness.

It is thanks to society that there is an intervention of consciousness. It is not thanks to unconsciousness that there is an intervention of society. It is thanks to society that there is the certainty of pre-prepared movements, domination of the conscious over emotion and unconsciousness. It is right that the French Navy is now to make it obligatory for its sailors to learn to swim.

From here we easily move on to much more philosophical problems.

I don't know whether you have paid attention to what our friend Marcel Granet has already pointed out in his great investigations into the techniques of Taoism, its techniques of the body, breathing techniques in particular.[23] I have studied the Sanskrit texts of Yoga enough to know that the same things occur in India. I believe precisely that by education, and at least at the bottom of all our mystical states there are techniques of the body which we have not studied, but which were perfectly studied by China and India, even in very remote periods. This socio-psycho-biological study should be made. I think that there are necessarily biological means of entering into 'communication with God'. Although in the end breath technique, etc., is only the basic aspect in India and China, I believe this technique is much more widespread. At any rate, on this point we have the methods to understand a great many facts which we have not understood hitherto. I even believe that all the recent discoveries in reflex therapy deserve our attention, ours, the sociologists', as well as that of biologists and psychologists ... much more competent than ourselves.

Notes

1. [In fact Sydney Holland. See Holland, 1902–3.]. NB. The notes in square brackets are provided by the translator, Ben Brewster.
2. Best (1924: 1, 408; cf. 135) [*sic* – the latter reference seems to be a mistake of Mauss's; could he have been referring to 1, 436 or 11,556, which refer to the gait of men and women respectively?].

3. Teichelmann and Schürmann (1840: 73); cit. Eyre (1845: 11, 241).
4. Hertz (1929 and 1960).
5. Even the latest editions of Ploss, *Das Weib* (Bartels's editions, etc.) leave something to be desired in this question. [See Ploss 1884; Ploss and Bartels 1905; Ploss, Bartels and Bartels 1935.]
6. [See Roth 1897: 182–83; Roth 1924: 693–96.]
7. Observations are beginning to be published on this point.
8. Ploss's large collection of facts, supplemented by Bartels, is satisfactory on this point. [see Ploss, Bartels and Bartels 1935: III, 183.]
9. Graebner (1923).
10. This is one of the fine observations from Graebner (1923).
11. Sachs (1933 and 1938).
12. [Sachs 1938 uses the terms 'close dance' and 'expanded dance'.]
13. [Sachs 1938: 59–61.]
14. I have just seen it in use at last (Spring 1935).
15. [Rattray 1923: 62–3, Figs. 8-12, 15–16.]
16. ['The kinematic elements of a machine are not employed singly, but always in pairs; or in other words ... the machine cannot so well be said to consist of elements as of pairs of elements (Elementenpaare). This particular manner of constitution forms a distinguishing characteristic of the machine'. Reuleaux, 1876: 43.]
17. [McGee 1898: 152. In fact the Seri live on the island of Tiburon and the adjacent mainland of Sonora province, Mexico, on the Gulf of California.]
18. [Krauss 1904–13; 1906–7; 1909–29.]
19. [Granet 1929 and 1930.]

References

Best, Elsdon. 1924. *The Maori*, Memoirs of the Polynesian Society, Volume V, Board of Maori Ethnological Research, Wellington, New Zealand, two volumes.

Eyre, Edward John. 1845. *Journals of Expeditions of Discovery into Central Australia and Overland from Adelaide to King George's Sound in the Years 18401; sent by the Colonists of South Australia with the Sanction and Support of the Government including an Account of the Manners and Customs of the Aborigines and the State of their Relations with Europeans*. London: T. & W. Boone, two volumes.

Graebner, Fritz. 1923. *Ethnologie*, in Paul Hinneberg et al., *Die Kultur den Gegenwart*, Part III Section 5. Leipzig: B.G. Teubner.

Granet, Marcel. 1929. *La civilisation chinoise, la vie publique et la vie privé*, L'évolution de l'humanité vol. 25. Paris: La Renaissance du Livre.

Granet, Marcel. 1930. *Chinese Civilization*, translated by Kathleen E. Innes and Mabel R. Brailsford. London: Kegan Paul.

Hertz, Robert. 1929. 'La prééminance de la main droite: étude sur la polarité religieuse' (1909), in *Mélanges de sociologie religieuse et de folklore*. Paris: Librairie Félix Alcan.

Hertz, Robert. 1960. 'The Preeminence of the Right Hand', *Death and the Right Hand*, translated by Rodney and Claudia Needham. London: Cohen and West, pp. 87–113.

Holland, Sydney. 1902–3. 'Swimming', *Encyclopaedia Britannica*, 10th edition (supplement to the 9th edition), Edinburgh, vol. XXXIII, pp. 140–41.

Krauss, Friedrich Saloman (ed.) 1904–13, *Anthropophyteia, jahrbücher fur folkloristische Erhebungen und Forschungen zur Entwicklungsgeschichte der geschlechtlichen Moral*, Leipzig, ten volumes.

Krauss, Friedrich Saloman. (ed.) 1906–7. *Historische Quellenschriften zum Studium der Anthropophyteia*, Leipzig, four volumes.
Krauss, Friedrich Saloman. (ed.) 1909–29. *Beiwerke zum Studium der Anthropophyteia*. Leipzig, nine volumes.
McGee, W.J. 1898. 'The Seri Indians', *Seventeenth Annual Report of the Bureau of American Ethnology to the Smithsonian Institution for the year 1895–6*, Part I. Washington, pp. 9–344.
Ploss, Hermann Heinrich. 1884. *Das Weib in der Natur and Völkerkunde, Anthropologische Studien*, Leipzig, two volumes.
Ploss, Hermann Heinrich and Bartels, Max. 1905. *Das Weib in der Natur und Völkerkunde*, 8th much-expanded edition. Leipzig.
Ploss, Hermann Heinrich, Bartels, Max and Bartels, Paul. 1935. *Woman: an Historical, Gynaecological and Anthropological Compendium*, edited and translated by Eric John Dingwall. London: Heinemann, three volumes.
Rattray, Robert Sutherland. 1923. *Ashanti*. Oxford: Clarendon Press.
Reuleaux, Franz. 1875. *Theoretische Kinematik, Grundzüge einer Theorie der Maschinenwesens*. Brunswick.
Reuleaux, Franz. 1876. *The Kinematics of Machinery, Outlines of a Theory of Machines*. London: Macmillan.
Roth, Walter Edmund. 1897. *Ethnological Studies among the NorthWestCentral Queensland Aborigines*. Brisbane, Australia: Edmund Gregory, Government Printer.
Roth, Walter Edmund. 1924. 'An Introductory Study of the Arts, Crafts, and Customs of the Guiana Indians', *38th Annual Report of the Bureau of American Ethnology to the Smithsonian Institution 1916–1917*. Washington D.C.: Government Printing Office, pp. 25–745.
Sachs, Curt. 1933. *Weltgeschichte des Tanzes*. Berlin: D. Reimer.
Sachs, Curt. 1938. *World History of the Dance*, translated by Bessie Schönberg. London: George Allen and Unwin.
Teichelmann, Christian Gottlieb and Schürmann, Clamor Wilhelm. 1840. *Outlines of a Grammar, Vocabulary, and Phraseology, of the Aboriginal Language of South Australia, spoken by the Natives in and for some distance around Adelaide*, published by the authors at the Native Location, Adelaide (Xerographic facsimile, South Australia Facsimile Editions no. 39, 1962).

6. 'Boy and two girls'. P.79.ACH1. New South Wales, Australia. Mounted Haddon Collection (CUMAA©). 'It is at this moment [adolescence] that they learn definitively the techniques of the body that they will retain for the whole of their adult lives' (Mauss, text 9).

Text 10

Technology (1935/1947)

M. Mauss, 1947, *Technology Manuel d'ethnographie.* Paris: Payot, second edition, 1967, pp. 29–83.

Translated by Dominique Lussier

This long and thorough discussion of technology appeared as chapter 4 of the 1947 *Manuel d'ethnographie*, a book composed of extensive notes and transcriptions of the lectures given by Mauss at the Institut d'Ethnologie and the Collège de France in the 1930s. Denise Paulme, who collated these notes and added much of the bibliography, also recorded that Mauss had put special emphasis on technical and aesthetic phenomena in his lecture course for 1935–36 (as he indeed mentions in his essay on the Techniques of the Body). Mauss had in fact been teaching technology since the late 1920s, as attested by other lecture notes (taken by Anatole Lewitsky, Jacques Soustelle, etc.) now held at the archives of the Musée de l'Homme. This 1947 publication had not been revised in any detail by its already ill author, and it confirms the impressions recalled by his students Levi-Strauss, Leroi-Gourhan, and Haudricourt. With his teaching equally designed to serve museographic research and as a fieldwork questionnaire for both professionals and colonial administrators, Mauss's message oscillated between telegraphic terseness and an urge for exhaustiveness. Inspiring but also frustrating to follow, the profound insights and novel perspectives conveyed through these lectures often threatened to be buried by a plethora of abstruse or impressionistic details.

The history of technology – of the study of techniques – is a recent one: the studies launched by the Encyclopaedists were not pursued by their successors.

The Pitt Rivers Museum in Oxford, the Horniman Museum in the London suburb, and the Cologne Museum offer excellent examples of the history of techniques.

Techniques are to be defined as *traditional actions combined in order to produce a mechanical, physical, or chemical effect, these actions being recognised to have that effect.*

It will sometimes be difficult to distinguish techniques from:

1. the arts and fine arts, since aesthetic activity and technical activity are on a par as regards creativity. In the plastic arts, no differentiation can be established other than the one that exists in the belief system of the artist;
2. religious efficacy. The difference lies entirely in how the native conceives of the efficacy. It is therefore necessary to estimate the respective proportions of technique and magical efficacy in the native's mind (e.g. poisoned arrows).

The combination of techniques constitutes industries and crafts. Techniques, industries and crafts, taken together, constitute the technical system of a society, which is essential to it. A correct observation of this system will have to respect the different proportions of its constituents.

Absolute precision is indispensable in the observation of techniques. The most insignificant tool should be named and located: who uses it, where was it found, how is it used and for what purpose, does it have a general purpose or a special one (e.g. the use of a knife)? It should be photographed as it is used, together with the object on which it is used, or with its end product; the photographs will show the various stages of production. The industrial system to which the object belongs should be identified; the study of a single tool normally implies the study of the craft as a whole.

Finally, the position of crafts in relation to one another conditions the state of society. The mistake of Karl Marx is that he believed the economy to condition techniques: the converse is true.

Investigation and collection will always go hand in hand. Duplicates are essential, since, for instance, the same fabric must be studied from several angles: weaving, spinning, embroidering, ornament, etc.

Investigation and classification can take various approaches. Starting from a logical viewpoint leads to the setting up of series, the study of types, and the study of style. The technological viewpoint will lead for instance to the study of the axe, but not to the study of all weapons indiscriminately. Lastly, the perspective of industry and craft will make possible a living description of the society: describing a dinner service will include the history of its manufacture and of the conditions under which it is used.

Techniques of the body[1]

Some techniques involve only the presence of the human body, but the actions they bring about are nonetheless traditional ones, tried and tested. The sum of bodily habits [*habitus du corps*] is a technique that is taught, and whose evolution has not yet come to an end. The technique of swimming improves day by day.

Body techniques should be studied with the aid of photography and if possible with slow-motion cinema.

The study of body techniques should be divided, according to age, into techniques dealing with:

childbirth (position during delivery, handling of the child at birth, cutting of the umbilical cord, attention given to the mother, etc.);

breast-feeding (posture of the breast-feeder, how the child is held).

Weaning is an important moment, which often signals the definitive physical separation of mother and child.

The study of *techniques among children* will include the study of the cradle, then of the whole of a child's life; education of sight and hearing, elimination of certain postures, imposition (or not) of ambidexterity, the study of the use of the left hand; finally, the deformations the child will undergo (deformation of the skull, scarifications, extraction of teeth, circumcision or excision, etc.).

Among adults, the following techniques should be studied in succession:

Rest while awake: standing, on one leg, lying down, on a bench in front of a table ...

Rest while asleep: standing; lying down on a bench; use of the pillow; of the head rest (apparently localised between 15° and 30° of latitude); use of the hammock.

The study of *bodily movements* should include *movements of the whole body*: do people crawl, do they walk on all fours? The gait will vary according to whether clothes are sewn or draped.

Breath and *breathing* differ while running, dancing or performing magic; the rhythm of breathing should be noted, together with the associated stretching of arms and legs.

Running will cover movements of the feet and arms, and the endurance of the runners. Study *dance*; and *jumping*: long jump, high jump, pole vault, etc. How does one take off? How does one climb [a tree]: with a belt, with spikes, or gripping with arms and legs? ...

Swimming is entirely determined by tradition. How do people start off, how do they dive? Do they swim using a board, or a beam?[2] Swimming races on a turtle's back are held all over the Pacific.

How do people carry out *movements applying force*? How do they push, pull, lift, and throw?

Note the *use of fingers and toes*; conjuring tricks and legerdemain (using the armpit, the cheek ...).

Gymnastics and *acrobatics* can be the object of detailed study.

In relation to the care of the body, note whether washing is done with or without soap (what is soap made of?). What are the procedures for excreting:

how do people spit, urinate and defecate? The study of perfumes and cosmetics, with the collection of catalogued samples, should not be left out.

The division of life according to the *timetable* followed by the natives will yield interesting results: some societies stay up late at night while others do not. Full moon nights are almost always festive nights.

Finally, *reproductive behaviour* should be studied, including the complications caused by deliberate mutilations,[3] and noting the presence or absence of sodomy, lesbianism and bestiality.

General techniques with general uses

Techniques, strictly speaking, are generally characterised by the presence of an instrument. 'Instrument' here includes all categories of instrument. The basic classification in this matter remains that of Reuleaux[4], who divides instruments as follows:

Tools. *The tool, which is usually conflated with the instrument, is always simple, consisting of a single piece of matter. Examples of tools are the cold chisel, wedge, lever.*

Instruments. An instrument consists of a combination of tools. An example is the axe, which in addition to the metal head includes a handle serving as lever; a knife fitted with a handle is an instrument, in contrast to a chisel; an arrow is an instrument.

Machines. A machine consists of a combination of instruments. For example, the bow comprises the wood of the bow, the string, and the arrow.

Starting from the Palaeolithic age, humanity can easily be divided up according to these various eras. Thus the Tasmanians did not know of axes, which were possessed by the Australians. This does not mean that the Tasmanians had remained entirely at the Chellean stage: they are Aurignacians, but without knowledge of the axe: their biface was hand held.[5]

The transition from tools to instruments, from the Lower Palaeolithic (Chellean and Acheulean periods) to the following eras, corresponds to one of the most significant upheavals to have rocked humanity.

The third era of humanity is the era of the machine, a combination of instruments. A bow, a trap, a boat with oars, such as the Eskimo *umiak*, are all of them machines. The Upper Palaeolithic is the great age of development of machines.

Finally, the sum total of techniques implied by the use of various machines converging towards the same goal constitutes an industry or craft: hunting implies the bows, traps and nets; fishing implies the boat and fishing tackle.

Some industries can reach an extraordinary level of complexity, e.g. the pharmacopoeia, or certain agricultural techniques: the use of poison is a sign of the perfecting of techniques, just as the preparation of manioc involves several processes to remove the poison.

Table 1 Classification of techniques

General techniques with a general use	physico-chemical (fire)	
	mechanical:	tool instrument machine
Special techniques with a general use or General industries with special uses[6]	basketry pottery rope manufacture and esparto goods glues and resins dyes and dressings	
Specialised industries with special uses[7]	consumption (cooking, drinks) simple acquisition (gathering, hunting, fishing)	
	production:	stockbreeding agriculture mineral industries
	protection and comfort:	dwelling dress
	transport and navigation pure techniques; science (medicine)	

Mechanical techniques

General principles of observation. Any object must be studied (a) in itself; (b) in relation to its users; (c) in relation to the whole of the system under scrutiny. The mode of manufacture will be the subject of extensive investigation: is the raw material local or not? Some calcites were transported over considerable distances; the search for flint deposits is typical of the entire Palaeolithic and Neolithic eras; several Australian tribes travel six hundred kilometres to fetch ochre. The same questions need to be asked if the object is made of softwood, or of hardwood. Sometimes too the tool is imported ready made. Study the different stages of manufacture, from the raw material to the end product. Then study similarly the mode of use and the production of each tool.

TOOLS. In the history of the beginnings of humanity no examples are known of humans who are wholly devoid of tools. *Sinanthropus* himself is found in association with a certain pre-Mousterian toolkit, linked with Chellean and pre-Chellean toolkits. Humankind enters history already equipped: as soon as humans appear, tools are there.

The main tools are classified as follows:

Weight and impact tools. One example is a club. The stick with a star-shaped stone typical of Polynesia is an instrument; and the Australian spear, cast with the aid of a thrower, is a machine. A Chellean pick is sometimes a tool of weight and impact, sometimes a point.

Friction tools: grattoir, scraper.
Piercing tools: knife, brace, etc.

INSTRUMENTS. The hammer is an instrument, and so is the axe. Since an instrument is a combination of two or several elements, each element should be studied separately, then in their relations.

Instruments can be distinguished as solid combinations (e.g. a knife fitted with a handle) and separable combinations (e.g. the mortar and pestle). A mill in fact consists of two mill-wheels; an anvil on its own is but half an instrument. These elements need first to be isolated, so that subsequently they can be brought together. A nail, a tenon, a mortise, a dowel are tools, but they form parts of instruments.

When the parts are not separate, the fundamental question is how to assemble them: this is where the greatest strength is needed but the least is found. Some assemblages can be made entirely of ropes and creepers: such is the case with Malayans when they are putting together structures; similarly, rope by itself is used by our carpenters for temporary scaffolding. Glues and resins are an important way of joining things: glue is used throughout Central Australia. The Moroccan region is very poor as regards junctions; the only methods known are glue and flour-and-water, and there are no good-quality joins in the woodwork. In Morocco, there are some poor-quality swing-ploughs built with a dowel and that is all. Such observations of detail can serve to characterise a whole civilisation.

The classification of instruments is more or less the same as that of tools, since the useful part of the instrument is a tool.

Crushing and pounding instruments. For instance, the axe (haft, head and junction). There are many ways of fitting the handle of an axe: the main one consists of a curved haft, strongly bound to the head with strings, but other forms are far more complicated: the junction involving a slit haft, or an elastic one. The adze is far better represented across the world than the axe proper: the entire Pacific region knows only the adze. A peak seems to have been reached with the Indian axe, which can be found in identical form, made in one piece, in Upper Dahomey (an observation by Graebner).

Scissors and pliers. The history of scissors is displayed at the Deutsches Museum of Nuremberg.

Instruments that resist forces. Anvil, mortar and pestle, dowels and tenons. Glues and resins (see below).

The file or grater can be a simple tool; or else it can be an instrument with complicated shapes; normally, the file is used on top of another file.

MACHINES. Together with the spear thrower, one of the most primitive machines – if the expression be allowed – is apparently the trap.[8] Elephant traps in India wholly defy imagination; the Eskimo traps, in which the fur of the animal must not be damaged, are extraordinary. The use of traps, with their interplay between springs, resetting devices, weights and balance implies the knowledge of a certain amount of mechanics; the knowledge remains unformulated, but exists nonetheless.

The bow is probably older than the trap. There are many ways in which the wood and string of a bow can be assembled; they can serve as the basis for a classification.

The simplest form of sling, namely the bolas, is in itself already very complex.

Moreover, a particular set of machines can on its own provide the basis for an industry, e.g. navigation. The boat, equipped with oars or paddles plied by humans, is a machine. Similarly, all scaffoldings are items of machinery. An important question here is that of devices used for lifting. Once a construction is set up, mechanics stop being dynamic and become static.

Sometimes, very complicated machines have developed where detailed work is required. The time element is irrelevant in the making of such machines, which were required for executing very delicate operations of fitting or threading, e.g. working with the brace drill to make coins; or the pump drill throughout the Pacific, especially among the Maoris.

Finally, all the general industries with a general use should be studied according to their raw materials: techniques using stone, wood (including paper), leather, bone, horn ... noting the relative importance of the various techniques in social life. Take account not only of what goes with what, but also of what is missing. It is hardly possible any longer to find in the French countryside a peasant who knows how to mend a metal vessel; villagers simply wait for Gypsies to pass through. A study of cultural areas, carried out only on the basis of arbitrary decisions and failing to take into account what is absent as well as what is present, is an incomplete and poor study.

Fire[9]

Fire is a significant means of protection; it not only produces heat, but also keeps off wild animals. For a very long period of time, fire must in the main have been conserved. So one should first of all study *methods for conserving fire*: firebrands made of twisted straw, torches, fire-covers. Fishermen in Concarneau still carry with them their fire horn, a cattle horn closed at one end and containing some smouldering fern or sawdust; the fisherman blows on the embers to light his pipe.

Location of the fire. The hearth is located sometimes in the doorway, sometimes in the middle of the house. In Tierra del Fuego the natives group their shelters around a shared fire. Chimneys appear only very late, since smoke is not generally felt to be unpleasant.

The study of the *methods of obtaining* fire is of considerable interest, since the discovery of these methods coincides with the appearance of the first machines. Fire can be obtained by friction, compression or percussion.

By *friction*, by systematically rubbing a piece of hardwood (male) in the groove of another piece of softwood (female). The friction can be carried out in various ways: *drilling, sawing, ploughing. Drilling* can be simple, with two cylindrical sticks sometimes handled by two people. The Indians produce fire by this method in less than twenty seconds. Drilling was the method used in India by the Brahmins to rekindle the sacred fire; in Rome it was the only means by which the fire of the Vestal Virgins could be lit. A refinement of drilling with two sticks is drilling with a *string*, practised in Madagascar and among the Eskimos; a string is wound and unwound in turn around the male stick. In the bow drill of the Eskimos a bowstring is wrapped around the male stick; the operator makes the bow move backwards and forwards horizontally, while holding the male stick in place with the aid of a bone cap held between his teeth. And lastly, the pump drill is known by the American Indians. In *sawing* – a method characteristic of Malaysia – the two sticks, male and female, are placed at right angles to each other; the rigid saw is often half a bamboo, while the flexible saw is a creeper. Finally, in *ploughing* – the only method known to the Polynesians – the male part is rubbed backwards and forwards along a groove in the female part.

Fire can also be obtained by *compression*, on the principle of the pneumatic lighter, whereby a piston supplied with tinder is driven violently inside a cylinder, and taken out immediately it catches fire. Europe discovered the pneumatic lighter at the start of the nineteenth century, but it was used much earlier in Indochina and Indonesia. Finally, *percussion* is known to peoples as miserable as the Ona of South America who strike two pieces of iron pyrites one against the other. Of course, the use of matches is very recent.

Study of the methods of obtaining fire should be followed up with study of the various kinds of tinder (willow catkin, birch moss, floss cotton, etc.).

The various *uses* of fire should then be observed. *Methods of heating*: red-hot stones thrown into a container of cold water; heating on shards; ovens (there are various types of oven; the oven is universally known to humankind, at least in the guise of the countryside oven). *Methods of lighting*: brazier, torches, lamps (made of stone, shell, pottery, iron ...). Fire can also be used to split stone; and to soften wood or to harden it. Finally, it plays a part in techniques of pottery and metallurgy.

Myths of the origin of fire[10]. In many societies, the blacksmith appears as the culture hero.

Forge and blacksmith. Blacksmiths, human beings who alongside the secret of fire possess the secret of the transformation of metals, are very often sorcerers and magicians; hence they stand apart in society. Throughout Black Africa, the blacksmith belongs to a distinct caste and is despised, but nevertheless he fulfils the role of a peacemaker. A collection of the various tools of the blacksmith (crucible, bellows, nozzles, etc.) should be supplemented with samples of crude ore, preserved in grease paper, with their exact location; ingots collected at the various stages of casting; and finally, the products of the forge (agricultural implements, arms, jewellery, etc.).

Each one of the diverse operations of modelling, wiredrawing, beating, tempering, soldering, patinating, burnishing, damascening, and niello work can provide scope for thorough inquiry. Distinctions will be made, where necessary, between bronze, copper, brass, tin, and zinc ware. The techniques of precious metals should be studied: according to Elliot Smith and Perry, gold is found everywhere traces exist of a megalithic industry.

The researcher should carefully gather all the myths concerning the different metals and their alloys (consider their influence on alchemy and metaphysics), as well as the traditional formulae of the blacksmiths.

Finally, the study of metallurgy should lead on to the study of industries using substitute materials – industries of wood and paper, stone, pottery, techniques using bone, horn ...

Instruments may be brought to a very high level of perfection. Among the best blacksmiths in the world are the Gold and other Siberian tribes. German metallurgy was far superior to Roman.

Special techniques with general uses/ General industries with special uses

Here, with the notion of division of labour, the notion of métier or craft specialisation begins to emerge: the French peasant is a general handyman, but craft specialisms exist.

As soon as there is a general technique with special uses there is division of labour: to operate instruments requires a skill that is not necessarily distributed evenly throughout society.

Generally, the division of labour is by sex or age. Naturally it is also by locality, depending on the presence of materials: a village of potters normally settles near a deposit of clay.[11]

The whole range of techniques is not distributed evenly across humanity: the finest basketwork comes from the Far East and America; the best hollow ware is found among the Annamites. Each art must be studied in itself, without considering whether or not it is primitive; the end products do not necessarily vary according to the quality of the equipment used: thus the delicate fabrics of Morocco are made on primitive looms.

Whatever the technique under scrutiny, *all* its products should be collected, and *all* stages in the making of the machine should be studied.

Basketry[12]

The highest points of basketry are not reached by Europe, and certainly not by France, where the best basket-makers are Gypsies. The finest basketwork in the world is found in the Far East and Central America, most especially among the Pueblo Indians. Excavations conducted among the cliff dwellers of Central America have yielded curious results: the finest basketry in the world is that found in the deepest archaeological layers.

The study of all basketwork should begin with the raw materials (give both the local and scientific names) and with all its forms, as well as the passage from the one form to another. Collect samples of the different stages.

For the principles of description one can use the instructions contained in the *Handbook of American Indians*[13]. There is a good classification in *Notes and Queries on Anthropology*, p. 245 ff.

Basketwork is made up of two series of elements which are repeatedly brought into contact. In *woven* basketry the two elements are intertwined as they are on a loom; but the materials used by the basket maker are relatively rigid and wide, for instance, leaves of the coconut palm or pandanus, or rods of willow or acacia. A second type, *coiled basketry*, is in fact sewn: on a frame of rods or grasses corresponding to the warp the craftsman makes stitches, using a bone or metal awl. Coiled basketry without a frame cannot be distinguished from a net.

Each one of these main types is subdivided into many categories; we shall list only a few forms of woven basketry. When each thread (or weft) crosses the standards (or warp) regularly, the result is *check work*; standards and threads are indistinguishable once the product is completed. When the weft elements cross at regular intervals more than one warp element, the technique is called *twilled work*; it lends itself to decorative combinations. *Wickerwork* differs from check work in that its warp is rigid. Lastly, in *twined work* two or several thread elements twine around the rigid standards.

Generally, production is entirely by hand, almost without the aid of instruments; it requires considerable dexterity. The movements of the hands should be described, photographed and recorded with a motion camera, but above all, make sketches. Samples of each type of basketwork should be taken at the three or four main stages of production.

All forms are derivations from elementary forms. The primitive form of the net is the thread, then come the various forms of braiding; plaiting or matting with three or four threads is a superior type of braiding.

The bottom of the basket often is the most difficult part. Is the basket conical or does it have a base? The bottom of a basket can be single, double or triple. Round baskets often have a square bottom made of four triangles joined edge to edge. Study the relationships between the geometrical shapes. From time immemorial a good number of the theorems in plane and solid geometry have been solved by women basket makers have without any need to formulate them consciously – basketry is often women's work.

After the bottom of the basket, one should study the upsett: how are the risers joined to the base? What of the fastener and, where present, the cover?

At each stage provide the ideology of the object; give the description in indigenous terms, including where appropriate the symbolism and mythology of each stage.

Lastly, study the decoration. The decoration comes from the presence of elements with different colours; the resulting effects can be striking.

One should then classify the various types of basketry.

Basket: all varieties of baskets, for each and every use: simple or double baskets; baskets with a handle; with or without supporting strings. The edge of the basket should be studied carefully. *The winnowing fan*. *Matting* plays an important role in some civilisations: the whole of the Pacific region has mats, and the whole of the East has carpets; the tale of the flying carpet exists wherever the mat is known. Finally, *waterproofed* basketwork, in which liquids are stored, serves as a transition towards pottery.

Basketry can also serve a variety of purposes: as sheaths of swords and even as armour in Micronesia, in part of northern Asia and in Northwest America. The hafting of the great stone axes of Micronesia is made of basketry. Basketry is used in addition to fashion plaits of hair, head-bands, bracelets, rings, and also ropes.

Consider the links between basketry and other crafts, especially pottery.

Next comes the question of *fabrics made of fibre*. Roofs made of thatch or of plaited palm fronds or the like can up to a point be regarded as basketry. The observations of Cushing in a short work entitled *Manual Concepts* – observations which deal above all with basketry as the foundation of geometry – have had a decisive role in the study of this topic.

Pottery[14]

Pottery appears to be less primitive than basketry, from which it may partly derive: in a great many cases, the mould of the pot is a piece of basketwork smeared with damp clay and left to dry under the sun in order to make it waterproof. Originally, pottery must have been a substitute on the one hand for basketry, and on the other hand for containers made of stone – the latter necessarily being very heavy.

From an archaeological standpoint pottery is the sign of the Neolithic, or at any rate of very advanced Upper Palaeolithic. Completely lacking in Australia and in Tierra del Fuego, pottery remains very poor in all the Pygmy countries. At present, there is no pottery in Polynesia proper; however, traces of its prior existence can be found in the region. It was probably abandoned partly owing to the influence of oven cooking, a mode of food preparation requiring no fireproof containers. In fact, one of the main purposes of pottery is to create containers for cooking food; waterproofed basketwork in some cases, hollow ware in other cases can substitute for pottery in a region lacking clay soil, even among peoples whose industries are in other respects advanced. One can still find some very primitive societies producing admirable pottery, for instance the Pima of South America with their huge amphorae.[15] Among the most beautiful pottery known is that of the Toukala of Morocco, made with a wheel identical to that of the Djerba in Tunisia, which is one of the most primitive known to humanity.

Pottery can be tested by the sound it makes.

The distribution of pottery is easily explained by deposits of clay. Trade in pottery exists almost everywhere, and over fairly long distances.

The investigation of pottery should start with an inventory of domestic and religious objects. A stroll in the market can yield unexpected results. Where does the pot come from? Who made it? Can the husband sell the vessel produced by his wife? See Malinowski's description of pottery markets in the Trobriand Islands.[16]

Production. Who makes the pots? Generally, there is specialisation by locality and gender. Very often the potter is a woman (as among the Kabyle).

Gather samples of earth; to keep the clay moist, wrap it up in damp rags and sticking plaster. Give the indigenous and the scientific name for the clay; where it came from; how it is prepared, and any admixtures. There are clay mines. The whole of South America possesses a myth about kaolin.

In studying the various types of pottery one should first consider pieces simply left to dry in the sun. Pisé is a form of pottery; the whole of Marrakech's fortifications is but a huge piece of pottery dried in the sun. The extent of this use of sun-dried brick across the world is considerable. Whole granaries can be nothing but simple pieces of pottery. All the terraced houses in Africa, Peru and Mexico are pottery.

Then comes pottery that is fired, either in the open air or in an oven.

Handmade pottery can be obtained by *slip casting*, *moulding*, or *pressing* the pottery body into moulds of fired clay. In each case the main difficulty is to move from the base to the edge, especially so when the object has a base with a particular shape.

In *casting* the craftsman uses an object such as a basket or gourd which he covers with clay, either on the inside or the outside; the mould may be made on purpose or not; it may be destroyed in the firing process or it may be reusable.

The artist moulding his pot starts with a single lump of clay to which he gives the desired shape either without using any instrument or using tools which are generally few and rather simple (mallet, bamboo knife, shell or piece of calabash used as smoother, etc.).

Finally, pressing is the most widespread method: the craftsman prepares slabs of clay that he curves before luting them together by applying pressure. Sometimes a single, long, spiralled slab will provide all the material for the whole pot. Traces of the processes of assembly are removed before firing.

Very often, casting, moulding, and pressing are used successively in producing a single pot.

The potter's *wheel* may originally have been a fixed support for the lump of clay which the craftsman pugged by making it turn between his fingers; this support (a flattish dish with notches on its edge) became a turning wheel and was fixed on a swivel (thus forming the 'whirler') prior to becoming the potter's wheel proper, simple or double, turned by hand or foot, either by the potter or by an assistant.

Throughout the course of the whole process, whether the craftsman uses a wheel or not, one should study the work of the fingers, the manual dexterity and, if the wheel is used, the work of the feet; notice the return strings.

Drying can take place in the open air or inside the house, in the sun or in the shade. Firing will be carried out with an open-air fire, in a hole in the ground dug for that purpose, or with a true oven. Notice the nature and arrangement of the fuel, and the means used to increase or decrease the amount of air reaching the fire or the pots, so as to modify the degree of moisture, etc.

Decorating is done before or after firing, or both before and after. It can rest entirely on the choice of clay body and the firing conditions (e.g. adding coal dust); it can be the result of burnishing, with a smoother made out of wood, horn, bone or shell, or the result of glazing; it can be the result of impressing, whether with finger or fingernail, with a string impressions from a rope (basketwork moulds explain the frequent occurrence of braids or checkwork as decorative motifs), with a piece of cloth or some other object, or with a stamp prepared in advance; it may result from an incision or an excision. Impressed or engraved decoration is sometimes heightened with the aid of white or coloured clays prior to firing or after it. Pottery can also be decorated by means of grooves, by ornaments applied on the surface; by applying a slip, that is a fine layer of white or coloured clay, or a glaze such as haematite, ochre or graphite; by varnishing with a resin or glue after firing; by applying lead glaze or opaque enamel, with or without colouring; by painting with coloured clay or with the use of other colorants ...

Use the philological method, asking for a description in local terms of the entire process of manufacture – including the knacks (tours-de-mains) – and of all the decoration.

Collecting. The classification of pottery is one of the most difficult sorts of classification.

Pots could be grouped, using the inventory made locally, in relation to their use. A second classification could distinguish according to form, size and decoration. Here we meet the notion of typology that will recur in the context of art. Some forms are very rare, the most difficult to realise being naturally cubic shapes. Note the myth and ideology associated with every shape and decorative motif. Technical features offer yet another sort of classification: a whole portion of the first wave of Celts possesses beaked vases; vases with or without handles, or bases. The study of decoration can serve as yet another basis for classification.

A complete collection should include all sets, and for each one of them, all the samples showing variations in type. Mention the range of size within which the particular type evolves.

For objects in high relief take an impression.

Symbolism of decoration: links with sculpture and modelling. Sculpture is the moulding of a volume.

Finally, study the relationships between pottery and other crafts.

There is nothing more uneven [in quality] than pottery, even in France; nothing, or almost nothing, is more variable. The *high points* of pottery are represented by the Central American complex, Peru and Northwest America, where wooden vessels are at the confluence between basketry vessels and pottery. In Africa, some peoples possess a very sophisticated pottery technique and terracotta sculpture; others only know of coarse pottery, barely fired. Nothing is more traditional than pottery, which is at once art and industry, and is felt as one of the most essential plastic arts.

Pottery made with the potter's wheel is not necessarily superior to that made without it; it all depends on the artist. And the perfection of the wheel does not necessarily correspond directly to the perfection of the pottery.

Pottery normally goes with an ideology. The question of the tripods, to take but one example, can be very complicated. Almost all pots have symbolic values; even in our cafés, a glass for port is not the same shape as a glass for a half-pint.

Very often the pot has a soul, the pot is a person. Pots are kept in a specific place, and they can often constitute objects of considerable religious importance. The Japanese *reku* vary according to season. Urns used as coffins are found in India, Africa and South America.

Finally, one should study the fate of each pot. What happens to shards?

Esparto goods and rope making

There is not much difference between basketry and rope making or between basketry and esparto goods. In the latter case the work is carried out with the whole reed or the whole leaf (this is the case in Papua and Melanesia), but the art of plaiting remains the same.

Underlying all fabrics one finds the notion of net and braid: a fabric is a net that has itself been netted.

Esparto goods can be exemplified by the manufacture of wicker shoes, or by the making of fabric out of dried leaves. Esparto work will produce more or less the same objects as basketry – e.g. the sheath for a sword.

For everything to do with plaiting it is always necessary to study the fibre; the strand, in other words a composite of fibres; and the thread, which can be a composite of several strands. To analyse such composites, employ the tool used in drapery, the linen tester, and count the number of threads per decimetre or square decimetre. The study of fibre and thread should be followed by the study of techniques for plaiting, for the weave, and for weaving.

Rope making differs from basketry in that it involves only the manufacture of the thread, i.e. the rope. There is the question of the twisting of the rope and its resistance to twisting; the casting off at the start and at the end. In the study of ropes the essential question is *knots*. One should unravel the interlacing of the various strands, noting the movements of fingers and hands that produce it. Knots are of considerable importance; the French can scarcely make a hundred knots while the Eskimos normally know at least two hundred.

Glues and resins

Glues and resins can be studied here as well as in connection with tools and instruments (see above). Glue, resin, wax and varnish are instruments that resist forces.

There is no published synthesis covering this topic. Glues and resins are much used in Australia. The existence of substances of this kind allows us to understand hypothetically the use of a certain number of prehistoric tools, whose function would otherwise be inexplicable.

One of the most efficient types of glue is blood. Among the types of glue do not forget *waxes* (beeswax and others). Study the various uses of wax.

Study the composition and modes of preservation of the *varnish* used in basketry, pottery, etc.

Weaponry[17]

Weaponry can be studied as forming a general industry with special uses – the same knife can be used for hunting, for war, for butchery; or else they can be studied according to their use: weapons of war, fishing weapons, hunting weapons. We can distinguish further between projectiles and impact weapons; and within the latter between crushing weapons and sharp weapons – cutting ones and piercing ones.

Table 2 Types of weaponry

Weight weaponry	Club (of wood, stone, or metal) Truncheon Hammer Biface Hunting spear Hoe
Offensive weaponry	Axe Knife and stiletto Sword (scimitar, yataghan, kris) Dagger
Projectile weaponry	Javelin, large and small Sling Bola Lasso Bow and arrow Blowpipe
Protective armour	Shield Helmet Armour
Parade armour	
Firearms	

Whatever the weapon under scrutiny, the investigation should cover, in sequence, the name; the raw material and the various stages of manufacture; its use, handling, mode of action, its range and effectiveness; who is entitled to use it (man or woman or both; is the weapon strictly personal or can it be lent to someone else, and if so to whom, etc.); finally, its ideology and its relations with religion and magic.

The inventory of the village's weapons carried out house by house and noting the name of the owner of each weapon will show the distribution of weaponry in the locality.

For each weapon one should note:

Its name: the general name and the individual one, if it exists – as indeed it often does; it is of some importance to know that Roland's sword was called Durandal.

Its myth.

Its material: an axe can be made of iron, stone, jade or obsidian.

Its shape: is it an axe proper, an adze, or a pickaxe?

The spear when held in the hand is a weight weapon; when thrown, it becomes a javelin. A removable head makes it a harpoon, which is thrown or held in hand and allowed to run out (whale hunting in the Torres Strait).

Offensive weapons

In a sword, dagger or knife the handle, guard and sheath need to be studied as much as the weapon itself. Roman swords were in fact Celtic swords: the best blacksmiths at Rome were Celts.

Projectiles

The study of the *bow* should include several steps:

1. The bow itself, its components and its production. The composite bow[18] is widespread, from the Mongol world to Central America; a composite bow can be made of between three and seven different kinds of wood. The cross-section of the bow: ellipsoidal or shaped like a lens. The Pygmies of the Philippines possess an enormous bow which is double in cross-section. The bow can present several different sections lengthwise.
2. The string and the way it is attached to the bow, whether fixed or movable. A very powerful bow requires a catch to which the string is attached when not in use. Study both the catch and the knot which fixes it in place.
3. The arrow (shaft, head, feathering, nock); is it poisoned (composition of the poison, its production, its effects, etc.)?
4. How is it shot? What is the posture of the archer, and the position of his fingers on the string?

One of the primitive forms of propulsion for projectiles is that provided by the spear-thrower, which is still in use in parts of Africa and America, and especially throughout Australia. The thrower is a small stick, about 50 centimetres long, with a hook; sometimes the hook is inserted in a cavity in the base of the spear ('male' thrower); sometimes the fluted stick ends with a concave butt on which the bottom of the spear rests ('female' thrower). A spear which, when cast by hand, has a range of twenty to thirty metres, can reach fifty to seventy metres when cast with a thrower. The thrower is often found at prehistoric sites.

Another primitive mode of throwing making use of a cord is throwing with a *sling*.

The *blowpipe* appears to be associated with the great equatorial forest. Its distribution is more or less that of the Pacific and American civilisations between latitudes 5° and 15°. The tube varies. It can be either a simple one or double (as for elephant hunting among the Sakai of Malacca, where the animal must be hit in the eye), or else it can be made of several segments with different calibres. The inside of the tube is either smooth or grooved; the arrows are nearly always poisoned.

Protective armour

Armour used for protection – armour, helmet, gauntlets, finger-guards, greaves – is also often used for parade.

The shield has a considerable history. The first shield was probably a simple stick enabling one to ward off blows from the enemy. Australia knows only of wooden shields, often very narrow. The Zulu of South Africa have an oblong leather shield, whose handle consists of a vertical stick. The leather disk of the shield moves on the axis of the stick; it turns at the slightest shock, thus deflecting the flight of the arrow. Study the history of shield handles.

A shield can be made of wood, skin, leather, metal or basketry. It can be round, oblong, oval or rectangular, or it can be a shoulder shield.

The shield usually is a personal weapon that cannot be lent. In a society that is even moderately warlike, the shield's decoration can indicate the exact rank of its owner. The decoration of a shield usually corresponds to the coat of arms. The great Kwakiutl copper shields in Northwest America are true escutcheons.

The entire ancient North America used to have *gorgets*, sometimes made of bronze.

The *helmet* is far less common than the shield and probably comes from the East. It is made of leather, basketry or metal.

Total body armour exists in Micronesia (in basketwork) and in Africa (in basketwork, leather, coat of mail in Chad; parade armour in padded cotton as in Niger, hiding both horse and rider).

Parade armoury

The finest weapons are for parades, for ostentation, e.g. the jade hatchet of New Caledonia.

In some cases parade weapons can constitute a currency – one so precious as to be used only on the occasion of solemn ritual exchanges, e.g. shields in Northwest America.

We know of two societies for whom the spear is truly the object of a cult: Rome and Black Africa.

Specialised industries with special uses

The industries examined so far – fire, basketry, pottery, etc. – suggest the idea of a series of techniques, that is of skills and tools adding up to a trade.

A pure technology (study of techniques), like that of Franz Reuleaux, has every right to limit itself to mechanical techniques – in fact, all other classifications only regroup elements already covered under mechanical techniques: a textile is nothing else but a system of resistance; the firing of a very fine enamel belongs among physico-chemical phenomena. This is a technology for engineers.

There is another approach to technology, that of the historian of civilisation. We have not only classified things in relation to the internal logic of mechanics, physics or chemistry; we have also grouped them according to the social contexts to which they correspond.

From this vantage point, an industry is defined as *an ensemble of techniques that combine towards the satisfaction of a need* – or more precisely towards the satisfaction of consumption. Needs are elastic in humankind, but it is the notion of consumption that allows us to define industries, i.e. systems of techniques appropriate for specific purposes, and the articulation of industries: thus hunting and fishing each form a system of general techniques with a general use, a system of general techniques with special uses, and a system of special techniques with special uses.

We shall now study techniques by classifying them according to the purpose they serve. From now on the technical aspects are no longer the only important ones, for the purpose in view determines even the technical aspects: fishing weapons differ from hunting weapons; and within fishing itself, trout fishing differs from gudgeon fishing.

Here we enter a field that does not belong to science alone, but one in which conscious practice is also involved. An inventor has a theoretical logic specific to himself; but it is this notion of the practical solution to a problem that is called the notion of the technician.

Under the word 'coordination' [*administration*] people too often conflate economics and techniques. Of course, in order for several techniques to combine towards a single purpose, everything must be mutually adapted. So there exists a category corresponding to the coordination of movements; within a single individual all the techniques are coordinated one in relation to the others. But a person is not merely an economic agent – *homo economicus*, he is also a technician. French peasants spend a great part of their time on bricolage, in other words on technical activities. Some populations demonstrate an astonishing degree of industry, which is completely lacking among close neighbours who are characterised by complete mental laziness. The latter will not even adopt instruments that are ready to use; they will borrow nothing and imitate nothing, simply out of clumsiness or nonchalance.

Approached as above, the study of techniques immediately raises several questions: division of labour, according to time, place, society, gender ...; the question of consumption and its relations with production; and finally, the relation between techniques and techno-morphology, i.e. the question of the location of industries and the question of trade, often over long distances.

Economic phenomena as a whole belong to this category, which the Germans call *wirtschaftliche Dimension*, but they belong only as superstructure, not as infrastructure[19].

We shall classify special industries with special uses starting with what is most material and close to the human body:

industry of consumption;
industry of acquisition;
industry of production;
industry of protection and comfort;
industry of transport.

Consumption

Fieldworkers too often fail to study the consumption of food.[20] This kind of work requires sustained attention. It should in fact cover at least a year: the basic diet, consumed in normal quantities during some months of the year, can sometimes be reduced to famine portions, for instance during the hungry season in agricultural societies. Yet again, the fieldworker should turn to making an inventory. He or she should note food consumption in several families representative of the society under scrutiny (rich, average or poor families), for instance during the last week of each month: the quantity of food and the modes of preparing it; who ate what? Study the relations between the cycle of consumption and that of production.[21]

Consumption is nearly always confined to the domestic sphere, that is to the family. Even in Papua where meals are taken communally, wives prepare and bring the dishes; so although people take their meals together, the cooking is still a family matter.

Meals

One should study each meal of the day, drawing up a complete inventory, including drinks. Who is eating? With whom? It is exceptional for men and women to eat together. Where do people eat? Give the times for the meals.

Nature of the dishes. The ingredients and their collection. The food that is eaten can come from far away, thus giving rise to extensive trade: salt in Africa; spices; some tribes in Central Australia send military expeditions to find a condiment called *pituri*, hundreds of miles away; the trade in maté; the spread of peyotl throughout Central America. On geophagy see the admirable work by Laufer.[22] As for cannibalism, distinguish between endocannibalism and exocannibalism: there are societies in Australia where it is customary to eat one's dead parents[23]; elsewhere, a victorious tribe will take slaves from among the peoples it has subdued and will eat them in the course of solemn ceremonies: this is still the norm among the Babinga in the Congo.

Order of dishes. This should be noted carefully. Such and such a morsel is normally reserved for such and such a member of the group.

Instruments of consumption. The fundamental instrument remains the hand. But which hand? And which finger? We can tell a Muslim easily by the fact that at a meal he strictly uses his right hand only, the use of the left hand for eating being prohibited. Forks are less common than knives; the first fork was probably one used by cannibals; the forks of man-eaters are often real artistic masterpieces (New Guinea forks). The spoon is used more frequently, yet it is not very widespread. Wooden plates are used across the whole of Northwest America. Note the use of a mat or a table, though the latter appears to be very rare.

Cooking

For any type of meat the mode of preparation should be studied from the moment the animal is killed until the moment the meat is eaten. One should proceed in similar fashion for each item of the meal: fish, cereals, green vegetables, etc.

Preparation of food. Study the mortar, the millstone, the mill, the methods of detoxification – e.g. of manioc. What kind of food is eaten raw, smoked or dried? For cooked food a distinction should be made between food that is boiled (the normal method of Chinese cooking, alongside frying), roasted (the oven is far more widespread than the spit), or fried. Note the substances used for cooking.

Preservation of food. Native people generally are much more far-sighted than is claimed. The Eskimos know quite well how to pass from one season to the next. Study their store-rooms and reserves buried in the ground. The Klamath of Oregon bury their seeds in the ground, adding a few leaves from a plant whose scent keeps off bears. Pemmican; smoked, dried and salted fish. In the Marquesas Islands, the breadfruit was stored in wells ten metres deep and five metres in diameter, walled with banana and coconut palm leaves. Such reserves could last fifty years. Every type of polar 'sauerkraut'.

Ideology of food

For each type of food study its relations with religion and magic, the connection with totemism, age, and sex; and the relationship with the dead and the living.

Prohibitions can be seasonal: a Jew cannot eat leavened bread at Passover. Prohibitions attached to a war expedition. Mention any food taboos and prejudices, taking care not to confuse religious prohibitions with mere rules of prudence. Above all, never forget that the needs to be satisfied are in the first instance social (in Australia, the food prohibitions imposed on the uninitiated boy leave him with nothing but a famine diet).

Condiments

The study of condiments is of particular importance. Learn all about the salt trade (in Africa), the pepper trade, the spice trade; the various types of oil or grease. Animal butter, vegetable butter (shea butter). Societies can be classified easily into those who eat their butter fresh and those who prefer it rancid, the latter being far more numerous. Study yeasts, ferments and sauces; and food that is left to go bad.

Drinks

The study undertaken for food needs to be repeated for drinks. Where, who, when, for whom, for what? Methods of drinking: with the hand, with a leaf, with a tube. Ideology attached to each type of drink, especially to fermented drinks.[24] The issues of purification, transport and preservation of the liquid. In the whole of Australia, big wooden plates are the only means known for transporting water. Transport is facilitated by the presence of gourds, calabashes and coconuts. Part of Australia lives by cutting the trunk of the gum tree.

The study of fermented drinks leads straight to religion. The issue of etiquette is most significant here: when does one drink, who drinks, etc. Millet beer; palm wine; rice alcohol, maté, chicha. The vine seems to be of Indo-Chinese origin.

Finally, the study of *narcotics* and *intoxicants*. Everything that is chewed: tobacco, betel nut, chewing-gum. Hemp, from which a drink is produced in Northwest America, and which is causing devastation in the Arab world; opium; tobacco – was it not preceded by something else in America? Lastly, one should collect myths concerning fermented drinks, and concerning all psychotropic substances.

Industries of acquisition

Simple acquisition (gathering, hunting, fishing) can be distinguished from production (stockbreeding, agriculture) in that it consists in the gathering of material things that will be used as they are, without further preparation. But the truth is that the distinction between acquisition and production is a secondary issue: the producer is never a creator, but only an administrator; human kind does not produce, it only manages production: to make a knife is not to create iron, but only to transform it through a succession of improvements. The Germans distinguish more aptly between *Sammler* and *Produzenter*.

From another point of view, we are accustomed to a division of humankind into three ages: beginning as hunter and fisherman, humans would have been stockbreeders prior to becoming sedentary at the stage of agriculture. Men of the Lower Palaeolithic would have been exclusively gatherers, hunters and fishermen, that is direct exploiters. I am not entirely sure of this. It seems that the beginnings of agriculture appeared very early on.

In the last analysis all of this needs to be seen as a matter of relative proportions: hunting and fishing are often found side by side with nascent or occasional agriculture. There is no opposition between herders and farmers but, more often, an exchange of produce. The individual Peul in West Africa can afford to confine himself to herding because he buys grain from his Black neighbours who are farmers.

Gathering

Simple collecting, or gathering (animal, vegetable), should be studied by making a collection of all the things the indigenous people gather, and by making a complete inventory of all that is gathered and all that is used. It is a serious mistake not to attach enough importance to natural production – on which human production rests.

The natives know very well what can be eaten, what can be drunk, and what is useful. They know the habits of insects and animals. A good study of gathering should go side by side with an investigation into ethnobotany and ethnozoology.

Gathering of animals. It is more widespread than is usually believed: dead quadrupeds, worms, caterpillars, snails, rats, bats, lizards, lice, and termites. Locust swarms. Gnat cakes in East Africa.

Gathering of plants. Europe's inventions in this regard are poor by comparison with those of America or Asia: 45 percent of Africa's cultivated species are of American origin.[25] The Australians are familiar with three hundred plants from which they eat the fruits, roots, and tubers.

The exploitation of the forest should be studied first: how do people climb? How do they get through thickets? How do they dig the earth to find tubers? As early as in Australia, women dig the earth with a sharp stick in order to find wild yams. The Babinga pygmies in the Congo dig up wild yams using a very long probe, one end of which is equipped with wooden boards attached with a creeper. They prepare the earth with the other end, and then thrust the cone into the soil; the earth piles up on the cone, which is then removed with a stick.[26] Gathering is well developed among the American Indians, and forms an essential part of their diet. The Californian Indians gather everything: groundnuts, berries, graminaceae, roots, bulbs, and especially acorns that they eat either boiled or roasted. East of the sierras, pine cones replace acorns; each tribe has its own pine zone, whose borders it should not cross. The Indians also dig up roots and tubers; the flour is eaten as a soup, as a stew, or as pancakes baked under ashes; while they are excellent basket makers, they have no knowledge of pottery. On the gathering of wild rice in the region of the Great Lakes, see the work of Jenks.[27] The Klamath Indians of Oregon pick the fruit of a nymphea called woka, in the marshes, from mid-August to the end of September; it is women who harvest these fruits, using canoes.

In Indochina, the search for camphor involves the use of a special language.

The search for rare oils, and rare gums.

Beekeeping. One of the most advanced forms was practised in Ancient Mexico. Who undertakes it?

Then come seafood and shells from the coast. Study the heaps of shells (*kjoekennmödding*), which form one of the most important elements for the study of the European Palaeolithic.

The equipment for gathering will include digging sticks, poles to knock down ripe fruits, sacks for gathering, back-baskets, etc.

Hunting

Hunting has its starting point in gathering: a given social group has its own hunting ground, and even if it is nomadic it will not hunt beyond its limits. The native knows his ground: waterholes, plants, nature, number and habits of the animals; finding himself elsewhere, he will very often feel lost. Hunting can be divided into small or individual, and large or collective (buffalo or elephant hunt; hunting with dogs in Europe; hunting with fire in Africa).

Hunting can be studied in two main ways: according to the weapon used, or according to the game hunted (weapons, technique, time of the year, etc.). An individual does not simply 'go hunting'; he goes hare-hunting, and not hunting just any old hare, but this particular hare which he knows well. So we should classify according to the people who hunt, the game being hunted, and the instrument used for hunting it.

Trapping is called passive hunting, since the hunter remains passive once the trap has been set; but the trap is a mechanical device which functions as such. All human groups know how to dig pits in which game can be made to fall; the Pygmies are said to know of no trap that is more elaborate; the Australians only have hoopnets or dams for fishing; the use of traps for big game requires mechanical ideas that go beyond the range of their thinking. Some Asian fences for hunting elephants, who are thereby directed towards a pit, reach huge dimensions. The great chicanes of the Iroquois for caribou hunting spread over tens of kilometres. The great nets of New Caledonia made the hair of the flying fox. Traps can be distinguished as follows.

traps which the animal can enter without injury, but from which it cannot escape (e.g. nets set horizontally or vertically; the lobster pot; hoopnets for fishing);

traps in which the animal is wounded and caught; these often include a bait that will lure the animal into triggering the trap mechanism (e.g. the mouse trap or elephant trap);

spring traps, where the spring can be triggered by traction (bird traps in Indochina) or by pressure (deer traps in Sumatra); the spring trap can also be set like a crossbow (rat traps in Madagascar);

birdlime traps, used in Hawaii to obtain the feathers of a bird, which is immediately released;

traps with a slip knot (parrot hunting among the Maori; squirrel hunting in Alaska; lark hunting in our own countryside; traps with radial sticks for hunting the antelope in Africa).

Whistle decoys allow the hunter to call game; they are often used in combination with a *disguise* that allows the hunter to come conveniently close to his prey. The Eskimos disguise themselves to hunt reindeer, the Bushmen to hunt ostriches (the hunter holds an ostrich head above his own head while imitating the bird's gait), and the Sudanese for hunting cranes. When hunting deer, the California Indians wear deer skins and fix antlers on their heads as they advance on all fours to windward of the game, pretending to graze; the Manchu Tartars behave similarly at the season of rut when stags look for each other in order to fight. The hunter imitates the game's call with the aid of a caller, and the deer runs forward presenting its chest to the hunter, who pierces it with his spear or sword.

The rules of hunting can vary depending on the type of game, the terrain, and the season.

The use of *dogs* in hunting; the pointer remains an auxiliary of secondary importance; only the hound can provide useful services.

The *ideology* of hunting, of the hunter, and of game. The hunter must know the names of the divinities associated with hunting and with the forest. He must be able to use incantations against the game and must know the meaning of omens. The whole of North America lives a life pervaded by the mythology of deer. An Australian hunter would not go hunting without holding a piece of quartz in his mouth.

The consumption of game generally has a ritual character and varies with the seasons.

The preservation of game, with its skull and bones. Do people break the bones and eat the marrow? Which parts can be roasted? Use of the remnants: fur, skin, guts…

Animals that are half hunted and half domesticated (e.g. pheasants) lead us to the question of domestication and stockbreeding. Pigs are half wild in Indochina, Melanesia, Papua, and Polynesia. Cattle pens. Goats were domesticated in pens.

Fishing[28]

Hunting and fishing haunt the mind; they occupy a great place in the preoccupations of the natives: the myth of the hunter and the myth of the fisherman are among the most important myths. Some of the usages and beliefs that get associated with totemism are in fact stories about hunting or fishing. The whole of Black Africa lives by the hunter; the whole of Melanesia conceives of its gods in the form of sharks.

From another point of view, fishing develops earlier than hunting. It is evidenced throughout southeast Australia by the presence of large-scale constructions – whole rivers provided with barrages.

Fishing should be studied just like hunting, according to the weapons used and the species targeted. Thus the trident, which is used very widely, is nevertheless adapted for each type of fish. Nets and traps are designed with a view to a particular kind of fish.

Bare hand fishing is practised by Tierra del Fuego women with the aid of a seaweed stalk weighed down with a stone; squatting in her canoe, the woman uses the flesh of a shellfish as bait, lets the line sink, and seizes in her hand the fish that comes to take the bait. Bare hand fishing is also practised in West Africa in the season of low water, in riverside backwaters that are then cut off; the whole village then gives itself over to fishing that is truly miraculous.

Net fishing is generally poorly studied. For each net note should be made of its manufacture (thread, method of weaving, size of mesh, use of the shuttle), its mode of use, and its setting. The net can be left in place, manipulated by hand, weighted, held up with floats, and can be used as draw net, drop net, casting net, trammel, etc.

Angling is done by hand or with a rod. Study all the elements of the fishing line: the line itself (a different thread is required for each fish); the hook, which can be composite (Polynesian hooks are among the finest); the bait and its fastening; floats and lead weights if present; sometimes, also, the decoy.

Angling and net fishing are relatively rare. Spear fishing is practised far more frequently. It is done with bow and arrow, with a gaff, or with a harpoon. The harpoon appears quite suddenly with the upper Palaeolithic. Some peoples are still unaware of such a weapon. How is the harpoon paid out? Spear fishing is generally practised from a fishing platform: standing on a kind of watch tower the harpooner pierces the bigger fish from fairly high above. The fishing platform is the same in part of Indochina, in Polynesia, in Papua, and throughout South America.

Trap fishing is significant for the diversity and the number of the types of trap observed: dams (simple, with an inlet, or V-shaped), dykes, large chicanes; all kinds of nets, often huge; fixed traps, with or without bait; mechanical traps, etc.

Finally, *fishing by poisoning the water*, practised in many African waterways, implies some subsequent procedure to detoxify the fish.

It would be impossible to exaggerate the significance of pearl fishing and tripang fishing in the history of the relations between the Oriental worlds of India and China, and the whole of Insulindia [southwest Pacific], and even Polynesia.

When practised from boats rather than from the shore, the study of fishing naturally involves the study of fishing boats; consider the relation with navigation.

Study reserves and dams. Study the legal rules of fishing.[29]

Fishing rituals can be very important. Australia has some very complex rituals, notably for calling up the whales which natives claim to be able to drive ashore.

Relations between fishing and social organisation. Generally, tribes are internally divided between moieties of fishermen and non-fishermen, the fishermen living on the coast and carrying on exchange with the non-fishermen who live further inland. Thus some villages are peopled by fishermen alone: here is one of the first forms of division of labour. Fishing involves an element of regularity that makes it easily susceptible to calculated exploitation.

Fishing can also be seasonal and depend on fish migrations. This applies to salmon fishing, which plays a major part in the life of the inhabitants of Northwest America. Accordingly, among fisherfolk, the migrations lead to the phenomena of double morphology affecting the whole group. In this case, the fishing villages, built for a few months of the year, will nonetheless involve the installation of fish tanks, drying areas, and warehouses to process fish.

The permanent villages of fishermen are often built on piles to withstand floods or storms. Conversely, building on piles or stilts may have been practised initially for purposes of defence; but it would then have led the inhabitants to practise fishing (lake dwellings in Switzerland).

Preparation and storage of fish. It may be eaten fresh, dried, high, rotten, and smoked.

Use of associated products. Fish oil, roe, condiments made of rotten and crushed fish, bladders ...

Industries of production

We have already seen that in fact there is no such thing as production by humans – only an administration of nature, an economy of nature: one breeds a pig, one does not create it.

Humans are animals who live in symbiosis with certain animal and vegetable species. They must follow their plants and animals. This explains the vast extent of migrations by certain peoples such as the Huns or the Peul.

However, while the industries of acquisition – gathering, hunting and fishing – correspond to direct exploitation, the industries we are about to study involve some alteration of nature, a difference that needs emphasis.

The study of a particular society necessarily includes study of the animals and plants of the society: the African elephant, half wild, was domesticated in antiquity; one does not know Dahomey without knowing the serpents that it worships. The stirrup was introduced to Europe in the eleventh century by the invasions of pastoral peoples from the East; previously, the horseman's weapon was inevitably the javelin, a projectile weapon, rather than the spear, a weight weapon. The entire history of Polynesian migrations is linked with the history of the plants and animals with which men and women set off in their ships and which they afterwards worshipped.

On the other hand, each animal, each plant has been worked on to an extraordinary degree. See the demonstration by Vavilov of the purely American origin of the two kinds of maize. We are only at the beginning of such creations.

The native has an acute awareness of the individuation of each animal and each plant: a Maori knows each sweet potato in his field and distinguishes it from all the others, just as a French gardener knows individually each one of his rosebushes. It is important that the observer acquire this idea of the individuation of each animal and each plant.[30]

Stockbreeding[31]

Stockbreeding scarcely appears until the Maglemosian, that is, the latest forms of the Palaeolithic. It arises suddenly together with pottery and brachycephalic humans. The latter brought to Europe pottery, stockbreeding, and agriculture.

Domestication may have appeared first on the slopes of the Himalayas. All domestic animals, or nearly all of them, come from this region.

The definition of the domesticated animal is an anthropocentric one: humans have domesticated the dog, but it is the cat who has domesticated humans. Moreover, some animals are tamed without necessity, for leisure (e.g. crickets in China).

The important issue in domestication is reproduction: certain species that are unable to breed in captivity remain half wild (e.g. the elephant, pheasant, Melanesian pig).

The investigation of stockbreeding should be made via the individual study of each domesticated animal, taken individually: age, sex, name, photographs, life history of the animal, naming of its body parts.

The *ethnozoology* of each species will include the study of its habitat, its origin (theories about the souls of breeding animals), and selection. There is a textbook on equine science by a Hittite prince dating from the seventeenth century BC. An Arab tribe possesses the pedigrees of its horses and is as proud of these as it is of its own genealogy. Researches on hybrids are often quite remarkable. At the courts of China, the Pharaoh and the Great Mogul everyone bred hybrids.

How are the animals fed? Fodder, grazing and migrations that result from the exhaustion of the grazing. Transhumance. Watering places.

How are the animals kept? Study of the pens. The kraal, an enclosure formed by huts built in a circle, with an open space in the middle where the animals are penned at night, is typical of the whole Bantu world. Study the shepherd and his relationship with the animals: the shepherd's call; his postures. Throughout East Africa, the shepherd rests on one leg like a wading bird. Has he given an individual name to each of his animals? Presence or (more common) absence of cowsheds.

Rearing. Castration; labour; delivery. What knowledge do the natives have of selection? How is each animal treated according to its age?

Use of the animal. As a means of transportation (see below). Is it eaten? If so, under what circumstances, which parts, and who has a right to it? Butchering is nearly always a sacrifice among the pastoral peoples of East Africa. The animals are killed with a spear or with an arrow shot at point-blank range. Use of the blood. The consumption of hot blood often takes a ritual character. Are the bones broken or are they not? What becomes of them (links with the ideology)? Is the marrow eaten? What is done with the gut, and skin? The working of skins is one of the most ancient industries known, as is witnessed by the large Chellean scrapers. Recipes for cooking. Do people make butter, and how? Do they make cheese, and how?

Veterinary arts are highly elaborated among the Sakalavas [of Madagascar]. Notions of pathology concerning the origins of disease among animals; therapy; surgery; obstetrics. The study of veterinary magic.

Ornamentation and deformation of animals. Example: pigs' tusks twisted in a spiral in Melanesia, Papua and the whole of Indonesia. Working with horn in Madagascar and throughout the Indian Ocean. Property marks on animals: who owns the mark, myth of the mark?[55] Take an impression of all the deformations, of all the marks; when these impressions are classified according to their owners, families or clans often become apparent.

Ideology. Study all the ceremonies connected with the worship of animals. Mythological and scientific notions on the origin of animals, theories about the souls of breeding animals, trophies of sacrificed animals. A Malagasy village is an ossuary of cattle skulls.

Law and economics. Animals serve as currency throughout East Africa. They were once the first currency of the Indo-European world (*pecunia* comes from *pecus*). On the other hand, the deforestation in North Africa, with all the changes it entails in the economy, coincides with the introduction of sheep to the region.

Agriculture[33]

Agriculture comes under ethnobotany, as stockbreeding comes under ethnozoology.

Agriculture is known throughout the Neolithic world. This technique is present in all the French colonies. Agriculture exists in nascent form among the tribes of northern central Australia. It is known to a good many Pygmies, notably those in the Philippines, and is apparently unknown today only to the inhabitants of Tierra del Fuego and the Arctic – the climate of these regions renders its practice impossible.

The theory that presents women as the sole inventors of agriculture seems to go too far. On the other hand, it is useless to ask whether agriculture does or does not represent a higher stage of civilisation than that represented by stockbreeding. Some purely pastoral civilisations have been great civilisations (e.g. the Mogul empire in the twelfth century). What matters rather is to know whether each breeder and each farmer is or is not of the highest standing within his technique.

Agricultural *implements* derive from the instruments used for gathering, and particularly from the digging stick, which started as a simple sharpened stick and evolved into a spade or hoe. Cultivation can fairly easily be divided into cultivation with the hoe, cultivation with the spade, and cultivation with the aid of primitive or evolved forms of plough. But the plough (and harrow) imply the use of domesticated animals, and hence the knowledge of stockbreeding. In many regions the plough has remained very primitive, a simple hoe that is dragged along. It is striking that the seeder, which is fairly widespread across the world, has been rediscovered in Europe only quite recently. Some prehistoric stone tools seem clearly to be primitive ploughshares.

The observer should distinguish as many types of agriculture as there are cultivated species.

He or she should study each plant in all its parts, at all ages from seed to fruit, noting the indigenous terms. The products from each plant can be very numerous and very varied. Some of these products rank among the most important that we use today; shea butter, palm oil, etc. are not European inventions.

The researcher should also trace the ideology that accompanies each plant – an important part of the study: the birth, life and death of the plant; its relation with vegetation, with Mother-earth, sky and rain.

One should study next the ecology and the economics of the plant: how the growing area is prepared, and how the plant is exposed to the elements. In some cases large-scale deforestation and clearing have only become possible with the introduction of metal tools. In the Stone Age the big obstacle for humanity was the forest, which could only be overcome by using fire; and even so the big roots would remain in place. Once the ground is ready it must be used: sowing, seedbeds, bedding out, final planting, maintenance. How is the ground irrigated (irrigation canals, wells, bailer, bucket water-wheel, etc.)? Manure is more or less universal. The fight against parasites. Then comes harvesting, threshing and storing. There is often a prohibition on storing two plants of different species in the same granary.

The study of *agrarian cults* should not be neglected. In countries with paddy fields the story of the rice spirit is fundamental in relation to rice growing, and not conversely.

When studying the techniques in themselves, it may be useful to distinguish: *agriculture*; *horticulture* (very often the garden corresponds to individual property, in contrast to the field which is collective property); *silviculture* (e.g. rubber, a wild plant maintained and cultivated in the forest by its owner); and *arboriculture* (e.g. coconut palms and olive trees). A good number of the inhabitants of our colonies are horticulturists even more than cultivators.

Study the sexual division of labour, marks of ownership and taboos.

The notion of productive surplus has been very well developed by Malinowski.[34]

One should also observe the relations between individual cultivation and collective cultivation, for the same patch of ground can be cultivated collectively as a field and individually as a garden, depending on the time of year; and the impact of these modes of cultivation on the social relationships both between individuals and between the individual and the clan as a whole.

Relations between agriculture and stockbreeding, between farmers and breeders, between farmers and hunting, and between farmers and hunters, if this is relevant.

Finally, some regions will need a study of quasi-industrial production: the Polynesian headman was a kind of general entrepreneur for agricultural labour.

Industries of protection and comfort

Protection and comfort should be analysed as conventional needs, far more than as natural needs. All the notions we have been developing since Adam Smith concerning the production of goods and the circulation of such products with a view to eventual consumption – all these are abstractions. The notion of production is particularly vague in its bearing on types of industry such as those that provide protection and comfort. It is vague not in relation to the notion of market but in relation to the idea of creation.

The elasticity of human needs is absolute: if we had to, we could live as Carthusian monks or nuns. With regard to protection and comfort, there is no other scale of values but the arbitrary choices of society.

Thus, beyond the limits of our civilisation, we straightaway find ourselves confronted with people having a scale of values, a mode of reasoning (*ratio*), and a way of calculating which are different from our own. What we call production in Paris is not necessarily production in Africa or among the Polynesians.

It is absurd that under the equatorial sun, a black Muslim wraps himself in as many robes as he can; but the accumulation of garments is the sign of his wealth. The garment is an object of aesthetic value as much as it is a means of protection.

The arbitrary character of everything that bears on protection and comfort is quite remarkable: the arbitrariness is not only 'economic', but in some aspects almost exclusively social. There can be *maxima* and *minima* of adaptation. Thus the Eskimos are perfectly equipped to fight the cold, as they are to fight the heat; their clothing is the sign of a very ancient Neo-Palaeolithic civilisation. The whole Arctic world is also very well provided for. But the Tierra del Fuego inhabitants are no better equipped for this struggle than were the Tasmanians, who have now vanished: both of these peoples went through very harsh winters with a miserable cloak of loose fur as their sole item of clothing.

It is therefore essential never to deduce anything a priori: observe, but draw no conclusions. If we want to be in a position to assess, we must first learn to beware of common sense, for there is nothing natural in this context. The human being is an animal who does reasonable things on the basis of unreasonable principles and who proceeds from sensible principles to accomplish things that are absurd. Nevertheless these absurd principles and this unreasonable behaviour are probably the starting point of great institutions. It is not in production properly speaking that society found its impetus to advance; but clothing is already a luxury, and luxury is the great promoter of civilisation. Civilisation always comes from the outside. Huge efforts have been realised in the area of techniques of production and comfort: the whole textile industry derives from clothing, and it is from the textile industry that a large part of the division of labour derives.

Clothing

Being an object of very slow consumption, clothing represents a real capital investment. It will serve as a means of protection while walking, running, attacking; it will protect against the bush, the rain, etc. Clothes worn during the day should be distinguished from clothes worn at night; working clothes from ceremonial garments, which are often found in greater numbers. The raw material will be determined by the environment, climate, etc.

In its protective function, clothing can be studied according to the part of the body it covers.

The shoe is fairly rare across the world. A great part of the world is without shoes; another part is very well shod. The origins of footwear appear to have been above all magical (for it avoids the foot being brought into direct contact with the ground and emanations from it), and military: the Australian *sandal* allows the tracks of attacking expeditions to be erased. All forms of the sandal should be studied: how do they hold the foot? Primitive forms of buttonholes. Sandals are footwear that is imperfect, inferior by far to the moccasin from which our own type of shoes derive; the moccasin is very close to Chinese footwear, and characterises the whole civilisation of Arctic Asia and central North America. The guild of shoemakers was probably one of the first to be constituted; its important role throughout Africa is well known.

The *gaiter* often comes in military forms (the greave).

In countries with palm trees, the body is protected by *rainwear*. Rainwear is more or less the same from its Asian centre as far as South America and part of North America. The study of sarongs, loincloths and grass skirts is difficult, but most useful. How is the waist covered? The way of wearing a loincloth and saluting with a loincloth by exposing the torso can by itself form a real language. On the other hand, the warlike character of a society may lead it to develop protective items for the torso: breastplates made of cotton, of basketry, and above all of leather, coats of mail ...

The *shirt* with stitched sides seems to be of fairly recent invention.

The *hat*, which is pretty commonly seen, is very unequally distributed. The Germans used to wear a small skullcap made of basketry while it does not appear that the Gauls covered their heads; only the military wore a helmet. Study all forms of skullcaps, turbans, helmets, broad-brimmed hats, etc. The wearing of a hat is often a sign of authority.

Moreover, some parts of the body may receive particular protection, for instance the penis (penis sheath, infibulations), and the body orifices – the latter for magical reasons.[35]

From the point of view of form, *draped clothes* should be distinguished from *sewn clothes*. Our buttons have no connection with antiquity whereas they can be found among the Eskimos and probably throughout the Arctic world; our buttonholes certainly are of Asiatic origin. Prior to their introduction, people only knew the fibula, i.e. the safety pin. But sewn clothes require the use of patterns, that is to say the vague awareness of a kind of descriptive geometry.

Decoration of clothes

Dyeing and decorations have played a great role in the development of clothes because of the quest for raw materials that they triggered. Raw materials had to be adapted to the desired dyes. The influence of fashion as regards dress seems to be immense. Clothes are one of the most reliable criteria for classification: thus Iroquois clothing is almost identical to Chinese. In this matter the influence of age, sex, era, etc. will also play their part.

According to the raw materials, clothing can be classified as follows.

Clothes made of skin, where the minimum is represented by people who wrap themselves in a single skin, floating freely, such as was current among the inhabitants of Tierra del Fuego; in contrast, Arctic clothing, though it too is made of skin, is entirely sewn. The tanning industry is very developed throughout the Arctic world, and in Sudan also; and the enormous quantity of scrapers found from the Chellean era corresponds in all likelihood to leatherworking. Study all forms of leatherwork, including leather bags, containers, etc.

Clothes made of leaves are found throughout Polynesia and Indonesia; the Kiwaï Papuans and the Marind Anim use a raincoat made of palm leaves.

The *beaten bark* of the fig tree, tapa, is used in Oceania and also in Black Africa. Some Indian fakirs are still dressed exclusively in clothes made from banyan roots.

Clothes made of straw can hardly be distinguished from those made of palm. Nearly all the clothes worn for masquerades in Melanesia and in Africa are made in this way.

Clothes made of esparto, plaits, fibres: those are the primitive forms of fabric, in terms both of material and use.

From fabrics in the strict sense one can distinguish *felt*, where the interlacing fibres are simply pressed, trodden and glued. Felting is known throughout north Asia and North America. It does not yield very impressive results, except in China and Tibet. Normally, felt is not very resistant; it tears, and it absorbs a lot of water.

Fabrics

Weaving is an important invention for humanity. The first woven material marked the beginning of a new era.

The study of any fabric implies the study of its *raw material*. Animal fabrics are made of wool (which is regarded as impure among the Egyptians); of goat or camel hair; horsehair yielded the hair shirt and crinoline. Silkworm breeding in China apparently goes back to the third millennium BC, but silk was only introduced to Greece by Alexander the Great, and to Rome by Caesar. Vegetable fabrics include linen, which was much appreciated by the Celts and the Germans; however in Europe, throughout the Middle Ages, its production was hindered by the cultivation of cereals. The taste for linen cloth only developed in our part of the world from the fifteenth century onwards. Other vegetable fabrics are hemp and above all cotton, whose history is not very clear: it was an Abyssinian plant that spread to India, but why it was not exploited in its country of origin while it became the basis for a fundamental industry in India, we do not know. Let us recall that the first cotton factories in England only date from the middle of the seventeenth century.

The study of a particular fabric presupposes the study of the *thread*, itself made of strands. The first raw material to be spun appears to have been hair. Human hair is spun throughout Australia. Notice whether the thread has one strand or several, whether it is twisted or twined; thread is twisted on the thigh everywhere in Burma. Study the way the fingers move, especially at the beginning and ending of the thread, which are the delicate moments; how do people prevent the thread from undoing? Study knots.

Study the spindle, the balance, the distaff. The spinning wheel appears later. Take photographs and, if possible, film in slow motion the way the fingers move.

Spinning can be extraordinarily delicate: for instance, the threads intended for making thin veils, which are found both in the Arab world and in the Egyptian and Hindu worlds.

After spinning comes *weaving*, in the strict sense.[36] Weaving is an industry that spreads across almost the whole of humanity, except those areas that lack the raw material (Polynesia, Melanesia, Australia). Some civilisations, now vanished, once had admirable fabrics.

To study weaving, one has to study the weaving instruments, the looms. If possible, collect some looms, noting carefully how the various parts are assembled. In studying the loom, the greatest care should be devoted to the moments where it is at a standstill during the operating process, and equally to each of the actions linking such moments. Every technical action ends with a point of stasis; it is a matter of describing how one reaches such a point and how one moves on from it to the next one. Each time, the relation between each action and each position of the loom has to be noted; the relation between every part of the weaver's body, notably his toes, and the loom. Take photographs, and above all make drawings, showing all the actions, and every step in the weaving.

Study the reels and the shuttles (shape, mode of throwing); in the absence of shuttles, what is the substitute?

When the loom is complicated and involves a handover between weavers, study the process of handover from one weaver to the next.

Note all the processes involved in making a warp and putting it under tension. The technique that consists in weighting the warp with pebbles disappeared in Norway only at the beginning of the nineteenth century; this is the usual method throughout the Arctic and American worlds.

Once the warp is set, the weft must go through it. How is the thread of the weft introduced and how does it leave the thread of the warp? By hand or with a sley? Procedure for stopping. Edging.

It is always difficult to achieve great widths; the very beautiful fabrics of Peru are made on very narrow looms. Techniques using ribbons.

Classification of fabrics: simple, twilled, plaited, combed, serge. In composite fabrics – e.g. tose using feathers, or those made on canvas – distinguish between the fabric that forms the background and that which forms the embroidery. A piece of velvet is ultimately comparable to a tapestry.

Starting from the second century BC, all the varieties of fabric had to be obtained from the Chinese and the Mongols. Thin veils and brocades reached the West, via Iran, only in the fourth century AD.

Dyes and dressings. When the thread has been dyed in advance, the decoration of the fabric relies on the colour of the thread. The dyeing of the thread should be distinguished from the dyeing of the whole fabric, where the resist method is often used (the dye will only reach a part of the fabric being worked on). Distinguish vegetable dyes from chemical dyes using mineral substances. Who does the dyeing? In Africa the dyer is often the wife of the shoemaker and, as such, she belongs to a caste.

The weaver. Who does the weaving? Where and when? Ideology of weaving and, if applicable, ideology of the animals who spin or weave (silkworms across the East, spiders ...). The Maori, like the Berbers, have a real cult of weaving.

Building[37]

As an essentially arbitrary phenomenon, the building of dwellings characterises a civilisation rather than a given territory. Architecture appears as the archetypal art, as creation par excellence. Building can then be studied among the industries of comfort and protection – not under human geography or the general history of civilisation, irrespective of the value of such studies. In the final analysis, building could be understood as a mode of consumption.

The researcher should not start by looking for the typical house: each house has its own sense. It is absurd to classify a society by a unique mode of dwelling; account must be taken of all the models found in the society, with all their variations, both individual and local: houses for general or special purposes, for human or for non-human use. Only when such a study has been completed can the notion of a typical house be abstracted without running the risk of confusing houses of rich and of poor men.

Types and materials

The house itself can be in a shelter. Numerous troglodytes or semi-troglodytes live in France: in the Cher valley, in Ferté-Milon, in Ferté-sous-Jouarre, etc. the inhabitants use old quarries. Caves were inhabited in Provence until the Bronze Age. Elsewhere, the civilisation of the troglodytes or *cliff dwellers* of central America (Arizona, part of New Mexico, north Mexico), raises a major problem in archaeology.

Houses made of clay can also form the end of the troglodytes' shelters; the house can be dug underneath, which creates a cave.

Simple distinctions between round houses and square houses seem to be inadequate. The same Gauls who had round houses built square granaries. So they were not incapable of conceiving of the two types. Throughout the north of France, in Flanders and Artois, farms made of bricks and cowsheds made of rammed clay can be seen side by side.

The simplest type of dwelling after the cave would seem to be the wind break: the Ona of Tierra del Fuego, as their sole protection against a harsh cold, make do with a screen made out of guanaco skins stretched over sticks driven obliquely into the ground in a semicircle around the fire. The Tasmanians knew of no other shelter, but they used strips of bark instead of skins.

The conical tent is widespread especially in the steppe regions. It is found throughout north Asia and in the north of America as far as Texas. The covering varies according to regions: in Siberia, deer skins give way further south to strips of birch bark, and then to pine and larch. The Plains dwellers use felt instead.[38]

In the *beehive style of hut*, the walls and the roof are not yet distinct. In the wigwam of the Atlantic coast Indians, poles driven in the ground have their tips bent and tied with transverse poles, themselves covered with grass, mats, or bark. The same type of building can be found with a circular or oval shape among the Pygmies of the Congo,[39] the Hottentots and the Zulus. The principle involved in building the Eskimo igloo, using ice, is entirely different.[40]

When the cylindrical frame of the walls is crowned with a separate conical roof, we have a dwelling of the same type as the Mongol tent covered with felt, or as the Siberian yurt with its roof of reindeer skins. The same type recurs in Africa with different materials: with walls of clay and a roof consisting of arches made of basketwork covered with grass, it looks like a large mushroom.

Finally, the *oblong house*, where the ridge beam is supported by several cleft stakes, can be found both in simple forms (e.g. in the Chaco) and with a complex architecture. Houses in British Colombia and throughout Oceania and Indonesia are made of wood with a double-pitched roof and protruding gable. The oblong houses of equatorial Africa have their walls made of bark, while the Muong of Indochina prefer basketry. Rammed clay walls often come with a terraced roof. Such is the case among the Hopi of North America, where the sandstone blocks are cemented with clay: access to the house is through the roof with the aid of ladders that can be withdrawn as a means of protection;

the terraced roof allows the building of one or several storeys set back; people go from one terrace to the other, from one house to the next via the terraces. The principle is the same as in Arabic architecture; nothing looks more like a Moroccan town with its terraced houses than a North American *pueblo*.

When studying dwellings, one should not overlook the possibility of a double morphology. One type of dwelling is not exclusive for a given area; the whole of north Asia lives in two ways: in a conical tent for part of the year, and in a round hut for the rest of the time.

It is probably normal, in some cases, that clay houses are built on stilts: this depends not only on the soil, whether it is sandy or clayey, but at least equally on technical factors. Sometimes too, a poor adaptation to the environment comes from the fact that people remain attached to a method that once had a raison d'être, but has ceased to be adequate because of a change in the mode of livelihood or because of migration. Moreover, old means can be adapted to new ends: in Oceania, the space between stilts serves as pigsty, and constructions on stilts provide excellent granaries. A number of large tombs serve in reality as hay cottages similar to those that are still to be seen in our mountains. From Indochina to France these cottages are placed on stone supports meant to prevent rats from climbing up; a detail of this kind can belong to a large-scale grouping of geographical and historical phenomena.

A thorough study of the different types of house found over a certain stretch of territory will make it possible to identify the limits of a given civilisation; but it is necessary to avoid the slightest a priori. Throughout southern Africa people camp in circles, the huts forming a circle around the hearth: this is the kraal. But there are people in southern Africa who no longer camp in circles and sometimes do not camp at all. So we should not say that all Bantus know the kraal, or that the kraal is specifically Bantu.

Functional and morphological study

To study the house, proceed like an architect: if possible, get local carpenters to make small-scale models of the various types of building. Each type needs at least three models, showing respectively the foundations, elevation and roofing. The essential task is to study the relations between the different parts.

The choice of location for a dwelling is often determined by reasons relating to magic or religion. The ground slopes, or has to be levelled, terraces have to be prepared, etc. Study the foundation rituals, which are often on a considerable scale; also the materials, and any excavation. Who does the building? In some societies, the carpenter is regularly the brother-in-law (wife's brother) of the owner. In Fiji, Melanesia and Polynesia, there are real carpenters' guilds. Note each moment, each detail and each action in the building. Deal with the techniques of building and the ideology of the techniques. Is building a collective undertaking or an individual one? Note everything to do with junction: tenons, knots, pins, nails, rafters, props. The balance of the building timbers. The ridge beam protrudes to form a canopy, or it does not. The shape of the roof can be more characteristic than the shape of the house itself. A house with round walls can be crowned with a square roof; but to put a round roof on top of a square house, as do the Bamoum of Cameroon, demands the solution of a difficult problem.

A house may well be built for a limited duration (e.g. the igloo of the Eskimos). The greater or lesser durability of the materials should not be the sole focus of attention: some houses made of dried mud can last for a very long time. However, one should note the concern for different kinds of shelter appropriate to the different stages of existence, depending on age, season, etc. Thus, dwellings for adolescents are common. Nearly all Negro families have their house and in addition a farm in the fields; very often the whole population spends the entire cultivation season away from the village, in the fields.

The model of ownership in Roman law is represented by ownership of land, and particularly of land with buildings. However, in nearly all African legal systems the house is regarded as the model of movable property.

In countries where the typical form of grouping is the joint or extended family the word 'house' will sometimes have to be applied to a group of dwellings. Consider a Norwegian farm: it includes one building for each purpose: the father's house, the sons' house, shelter for the forge, the stable, pigsty, cowshed, barn, kitchen, granaries, tool shed ... The whole constitutes a single house. The same is true of the Maori house and the Sudanese house.

The *plan* of the house will show its orientation, which is often very important: the Betsileo house serves as sundial.[41] Who lives in each corner? Give the location of everyone and of every object; also the space reserved for guests. The detailed study of furniture (how it is used, beliefs related to each object) should be accompanied by a plan with a scale. The kitchen, the hearth and the fire. How does the smoke exit (chimneys only appear rather late)?Beliefs and practices relating to the domestic fire. Relations between the house and the garden, the house and the fields; systems of enclosure. Surrounding walls made of stone are rare. Fences, hedges, etc.

Maintenance of the house. Is it resurfaced at fixed intervals? By whom and when? Where are the dead located, if this is relevant? Destruction of the house, e.g. in the case of death.

The study of ornamentation implies the study of every detail of the house. The owner of the house should be asked for explanation of every ornament. Such questioning will often reveal the use of blazons, and hence the presence of an aristocracy.

Purpose of buildings

Besides houses for private use, study with special attention houses for public purposes, particularly men's houses; if they are present, the houses of secret societies, which can be indistinguishable from temples. All Papuan villages are divided into two moieties, each of which has its own men's house. Elsewhere, the building of a men's house is an act no less important than the erection of a royal palace. It can be the sign of the emancipation of those who build it, it can provoke a war, and lead to sacrifices – or to the building of other houses.

Houses of the patriarch, of his wives, his daughters, his married sons, of adolescents.

The house of menstruating women.

Granaries, workshops, etc.

When such a study has been completed, the researcher can turn to the geography of technology. Once the various types of dwelling have been established statistically, one should abstract the notion of the architectural canon. But here, as with all social phenomena, the canon, i.e. the rule or ideal, should be distinguished from the average that is observed.

Agglomeration

The house normally has no independent existence, except in countries where the settlement pattern is essentially dispersed, but this is rare. A study of the dwelling would not be complete without a study of the village or, where relevant, the town. This issue, too often studied in purely geographical terms, should in my view be dealt with at least as much in terms of statistics and techniques.

The village is very often fortified, or else it is built in a military location. The location of a Ligurian village, a Kabyle or Maori village, a Betsileo village is the same: built on a spur allowing the inhabitants to dominate the landscape, the agglomeration is generally accessible only from one side. This is the case with the ancient [Celtic] *oppida*.

The location of a village is not necessarily fixed once and for all: the towns of the Gauls were alternatively located in the plain or on hill tops, and were sometimes fortified, sometimes not.

Study temporary camps and caves where people seek shelter; wells; all the collective services. The location of the refuse heap may be determined by religious factors. Use of refuse.

One should then study the village location from a geographical perspective, in relation to fields, to means of transportation, to roads and bridges. The study of towns is a major topic in the history of civilisation.

Some towns are fortified, side by side with towns that are defenceless. The fortified town in this case may correspond to an imperial or royal town. Thus the king's town stands out clearly in Mossi country and as far as the coast of Guinea.

Transportation industries[42]

Transportation industries are far more developed than is usually believed. From a technical viewpoint, the world became populated thanks to means of transport: just as the Sahara and Arabia are only habitable for camel breeders, certain parts of South America are only accessible to Indians who know how to use boats. Nevertheless, there exist societies which are still very poor in this respect: small nets are all that Australians possess to carry things, while the whole of north Asia possesses chests, which are often of large size.

Communication routes

First of all one should observe the layout of the ground: tracks, paths and roads. Show the indigenous names on a map. The use of an airplane will be very helpful for this in savannah or desert landscapes. Tracks can sometimes be seen from the air, even in forest. Relations between villages can exist over very long distances, and relations between tribes are not rare. If appropriate, note the techniques for protecting paths, typical of the whole of Indochina; the presence of barriers, fortifications, chevaux-de-frise; sometimes also, protection is ensured by religious prohibitions and taboos.

Bridges seem to be even more common than roads: rope bridges, bridges made of creepers, in Africa and America; suspension bridges throughout Asia, Oceania, and South America. Apaches in North America used to carry their horses from one edge of a canyon to the other using transporter bridges.

Porterage

On all these routes one has to carry. *Man*, or more generally woman, was the first beast of burden. The man holds the spear and the shield; if he carried the burden, he could not protect his wife.

How do people carry things? What instruments are used? The methods for distributing the weight on the body should be observed carefully. One of the reasons for positing the kinship of Asia and part of Oceania with America is the practice of porterage with a headband. Tibetan caravans descend to Nepal, climb the Himalayas and descend again into India carrying everything on the head with the aid of a headband across the forehead. Porterage on the head; with a shoulder yoke; on the hip. Study each time the gait of the porter, especially on uneven ground.

Next come *vehicles*. The first means of transport on land was no doubt of the trolley type: two poles are dragged along with their tips joined together and taking the load at their centre of gravity. Under the name *travoy* (from the old French *travois*), it can still be seen in North America.

The sledge is probably either prehistoric, or at least as old as the present north Asian civilisation. The Eskimo sledge remains the best.

The wheelbarrow, which implies the wheel, is very ancient throughout Asia. In the Chinese wheelbarrow the wheel was inside the tub. The theory of Mason and Powell on the origin of the cart seems correct: it proposes a trolley pulled by two horses with a wheel added at the centre of gravity. It is a remarkable thing that the Amerindians are aware of discs, and in addition they know the trolley, yet they never equipped a travoy with wheels. We can therefore assume that when they came to America they knew the travoy and the wheel, but did not have the idea of putting the one on the other.

The use of animals for transportation (*saddle animals, packsaddle animals, draught animals*) greatly modified the situation.

Here, the same questions arise as with human porterage: how are things carried? What is carried? Etc.

The reindeer must have been a beast of burden in the fairly distant past, but elsewhere than in America. The Eskimos followed the wild reindeer, but they did not domesticate it. Even those Eskimos who live side by side with North American Indians, who possessed domesticated reindeer, were not able to tame this animal; nevertheless, the Eskimos belong to a civilisation of the reindeer.[43]

Beasts of burden exist almost everywhere. Depending on the regions, we find the llama, the yak and the horse. The South American llama that yields the vicuña belongs to the same family as the Tibetan Lama. The arrival of the horse in North America transformed the whole country.[44]

In everything to do with draught animals, study the harnessing and yoking. In antiquity, the most developed yokes were the Asian ones, and among them, those of the Mongols.

The history of the cart and the wheel is among the most important of all historical developments.

When observing the mode of transportation, do not fail to mention how the animals are looked after.

Water transport

Water has never been an obstacle; water is a means of transport. Rivers have never been a hindrance to trade; they made it easier.

What sort of water is being crossed? Lakes, rivers, lagoons, and especially the sea, each require an appropriate means of transport.

The most primitive form of transport on water is no doubt represented by floating wood. In all the lagoons along the coast of Guinea the natives keep afloat by holding on to pieces of wood.

Several trunks joined together will form a raft, in the simple form still in use among the natives of the Amazon. The Maricopa Indians use only two parallel trunks, tied together crosswise with sticks; they propel the craft using long poles.

Sheaves of reed or rush tied together in the shape of a cigar are called balsas. Like the raft, the balsa floats on account of its specific gravity, but it is not watertight. It was known even to the Tasmanians, and it is still used on the Tchad and on South American lakes.

Inflated skins used as floats were the starting point for the development of boats made of skin: the kayak and the umiak of the Eskimos, the round boat [coracle] in Ireland, the latter made of an ox skin stretched across a skeleton of branches shaped into a hemisphere.

Canoes made of bark (Canada, Guyana) are so light that they can be carried when it is necessary to bypass cataracts.

But the most widespread of all primitive boats is no doubt the dugout made from a simple tree trunk hollowed out with the aid of an adze and fire: it goes back to the Swiss Neolithic and is common to Africa, the two Americas and Oceania. Such a boat can only be built when suitable timber is available. In Oceania the natives possessed simultaneously single-trunk canoes having their sides raised with planks, and double canoes [catamarans], which could be separated or tied together in a durable manner. The Melanesians travel on their rivers in simple dugouts made of a hollowed trunk which they propel with a paddle, but for the sea they use a boat equipped with an outrigger, which runs parallel with the boat and provides balance. The outrigger is typical of Polynesia and Indonesia, whence it spread as far as Madagascar. The large boats in Oceania measure more than thirty metres in length; the double canoes in Fiji used to carry a hundred passengers and several tons of cargo.

The Polynesians are admirable navigators. Their homeland is probably somewhere in southern Asia; in the course of time, they swarmed across the whole Pacific and as far as Easter Island.

The study of boats can scarcely be undertaken except by a sailor; he or she should study the planking, the stern and the bow. The keel is a recent invention, dating back barely to the seventh or eight century AD. The invention of the stem rudder is necessarily a recent one since it presupposes the presence of the keel. It is the Normans who brought in this development between the ninth and the twelfth century. The invention of the stem rudder changed the entire art of navigation.

The study of the boat's ornamentation will always yield interesting results: the boat is an animate being, the boat sees, the boat feels. Very often it has an eye[45], sometimes a neck, often teeth, whence the name of the Norwegian drakkar (dragon): it bites. Melanesian, Polynesian and Papuan boats have teeth.

The boat is a machine driven by an engine, with the aid of a particular mode of transmission. The simplest mode of transmission will be that provided by pole, paddles, scull or oars. Study the synchronisation of the paddlers or rowers, and the songs of the paddlers; note all the beliefs and all the rites concerning paddling. It is curious to note that the North American Indians who live in contact with the Eskimos never learned from them how to use the oars that Eskimo women use on their large umiaks.

The sail was a great invention. The triangular sail, at first without a mast, is known throughout the Pacific, while the Chinese junk is equipped with a square sail, generally plaited. Note all the systems of ties, all the knots, and the sails.

How do people orient themselves? The stars as points of reference. Do the natives know how to take a bearing? Do they have maps? Etc.

Living on boats, and boat houses.

Notes

1. On this topic see M. Mauss 1935, 'Les Techniques du corps' *Journal de Psychologie*, 1935, 32, pp. 271–93.
2. [*poutre:* mistake for *outre* (animal skin inflated and sewn up)?]
3. A. de Villeneuve, 'Etude sur une coutume somalie: les femmes cousues', *Journal de la Société des africanistes*, VII, fasc. 1, 1937, pp. 15–32.
4. F. Reuleaux, see especially *Theoretische Kinematik*, Berlin, 1875; *Der Konstrukteur*, Berlin, 1895.
5. Noting the lack of knifes among the Seri Indians, MacGee concludes that this is a society with a 'primitive' character (*The Seri Indians*, Bureau of American Ethnology, 17th Annual Report, 1895–1896); the comment is inadequate.
6. This heading covers numerous crafts, which however are well-defined and follow procedures and forms imposed by tradition.
7. All industries imply a division of labour; firstly in time, if the tasks are performed by a single individual, and then between the workers, who each take up a specialization.
8. On traps in general, E. Mérite, *Les Pièges*, Paris, Payot, 1942. There is a good study of traps, with a description in native language for each type, in F. Boas, *Ethnology of the Kwakiutl*, United States Bureau of Ethnology, 35th Annual Report, 1913–1914; and *The Kwakiutl of Vancouver Island*, Memoirs of the American Museum of Natural History, The Jesup North Pacific Expedition, vol. V, 2, 1909, pp. 301–522.
9. W. Cline, *Mining and Metallurgy in Negro Africa*, American Anthropologist, General Series of Anthropology, no 5, 1937. W. Hough, *Fire as an Agent in Human Culture*, Smithsonian Institution, US National Museum, Bulletin 39, Washington, 1926. A. Leroi-Gourhan, *L'Homme et la matière*, Paris, 1943, pp. 202–13.
10. Sir J.G. Frazer, *Mythes sur l'origine du feu*, Paris, Payot, 1931.
11. The lack of pottery in a large part of the Pacific has led some to posit the primitive character of the civilisation in question, without taking into account the lack of clay in the region concerned.
12. H. Bobart, *Basket Work through the Ages*, London, 1936. F. Graebner, *Gewirkte Tascher und Spiralwulstkörbe in der Süd-see*, Ethnologica, II, 1. H.K. Haeberlin, J. Teit and H.H. Roberts, under the direction of F. Boas, *Coiled Basketry in British Columbia and Surrounding Region*, Smithsonian Institution Bureau of American Ethnology, 41st Annual Report, 1919–1924, pp. 119–484. A. Kroeber, *Basket Designs of the Indians of North West California*, Berkeley, 1905. G. W. James, *Indian Basketry*, 2nd edn., Pasadena, 1902. A. Leroi-Gourhan, *L'Homme et la matière*, Paris, 1943, pp. 284–89. O.T. Mason, *Aboriginal American Basketry...*, US National Museum Report, 1901–1902 (1904), pp. 171–548.
13. *Handbook of American Indians ...*, edited by Frederick Webb Hodge, Smithsonian Institution, Bureau of American Ethnology, bulletin 30.
14. J. Déchelette, *Manuel d'archéologie préhistorique, celtique et gallo-romaine*, Paris, 1924–1929. L. Franchet, *Céramique primitive*, Paris, 1911. F.W. Hodge, *Handbook of American Indians*, Smithsonian Institution, Bureau of American Ethnology, bulletin 30. W.H. Holmes, *Aboriginal Pottery of the Eastern United States*, 20th Annual Report of the Bureau of American Ethnology, 1898–1899, Washington, 1903. A. Leroi-Gourhan, *L'Homme et la matière*, Paris, 1943, pp. 218–35. R.H. Lowie, *Manuel d'anthropologie culturelle*, French edition, Paris, Payot, 1936, p. 147–56.
15. J. MacGee, *The Seri Indians*, Bureau of American Ethnology, 17th Annual Report, 1895–1896.
16. B. Malinowski, *Argonauts of the Western Pacific*, London, 1922.
17. H.S. Harrison, *Handbook of the Horniman Museum. War and the Chase*, London, 1924. A. Leroi-Gourhan, *Milieu et techniques*, Paris, 1945, pp. 13–68. R.H. Lowie, *Manuel d'anthropologie culturelle*, French translation, Paris, 1936, pp. 232–42. G. Montandon, *L'Ologénèse culturelle. Traité d'ethnologie culturelle*, Paris, 1934, pp. 368–495 (to be used only with caution).
18. H. Balfour, 'On the Structure and Affinities of the Composite Bow', *Anthropological Institute Journal*, 1900.
19. On this point, see P. von Lilienfeld, *Gedanken über die Sozial Wissenschaft der Zukunft*, Milan, 1873–1883.

20. See the works carried out under the direction of Malinowski, notably: H. and S. Fortes, *Food in the Domestic Economy of the Tallensi* (Gold Coast), *Africa* IX, 1936, pp. 237–76. M. Hunter, *Reaction to Conquest*, Oxford, 1936. A.S. Richards, *Hunger and Work in a Savage Tribe*, London, 1932; *Land, Labour and Diet in Northern Rhodesia*, Oxford, 1940.
21. For planning the investigation, see R. Firth, 'The Sociological Study of Native Diet', *Africa*, VII, 1934, pp. 401–14.
22. B. Laufer, *Geophagy*, Field Museum of Natural History, Anthrop. Ser. Vol. 18, no 2, Chicago 1930.
23. S.R. Steinmetz, 'Endokannibalismus' in *Gesammelte kleinere Schriften zur Ethnologie und Soziologie*, t. I, pp. 132–260, Groningen, 1928. H. Kern, *Menschenfleisch als Arznei*, Ethnographische Beiträge, Festgruss 3. Feier des toten. Geburts A. Bastian (suppl. Int. Archiv f. Ethn. pp. 37–40). Th. Koch, *Die Anthropophagie des Süd-Amerikanischen Indianer*, Int. Archiv f. Ethn., 1899, XII, 2–3, pp. 78–111. E. Volhard, *Kannibalismus*, Stuttgart, 1939.
24. Ph. De Felice, *Poisons sacrés, ivresses divines*, Paris, 1936. C. Lumholtz, *The Symbolism of the Huichol Indians*, Memoirs of the American Museum of Natural History, v. III, Anthrop. II, The Jesup North Pacific Expedition, 1898; *Unknown Mexico*, London, 1903. A. Rouhier, *Le Peyotl*, Paris, 1927. T.T. Watermann, *The Religious Practices of the Dieguero Indians*, University of California, publ. in *American Archaeology and Ethnology*, 1910, VIII, 6, pp. 271–358. For the Indo-European world: G. Dumézil, *Le Festin d'immortalité*, Annales du Musée Guimet, Bibl. Et., t. XXXIV, Paris, 1934.
25. See A. de Candolle, *L'Origine des plantes cultivées*, Paris, 1883. A.G. Haudricourt and L. Hédin, *L'Homme et les plantes cultivées*, Paris, 1943. P. George, *Géographie agricole du monde*, Paris, 1946.
26. G. Bruel, 'Les Babinga', *Revue d'ethnographie et de sociologie*, 1910, pp. 111–25.
27. A.E. Jenks, *The Wild Rice Gatherers of the Upper Lakes*, 19th Annual Report of the Bureau of American Ethnology, 1897–1898, II, pp. 1019–1137.
28. On fishing, see E. Best, *The Maori*, Wellington, 1924, 2 vols. H.L. Roth, *The Natives of Sarawak and British North Borneo*, 1896, 2 vols. Th. Monod, *L'Industrie des pêches au Cameroun*, Paris, 1929. O.T. Mason, *Aboriginal American Harpoons*, Smithsonian Institution Report 1900, Washington, 1902.
29. On the rules of fishing: R.S. Rattray, *Ashanti*, Oxford, 1923.
30. See M. Leenhardt, *Gens de la Grande Terre*, Paris, 1937.
31. O. Antonius, *Grundzüge einer Stammgeschichte der Haustiere*, Iena, 1922. I. Geoffroy Saint-Hilaire, *Domestication et naturalisation des animaux utiles*, rapport général à M. le Ministre de l'Agriculture, Paris, 1834. E. Hahn, *Die Haustiere...*, Leipzig, 1896. B. Laufer, *Sino-iranica...*, Chicago, 1919. A. Leroi-Gourhan, *Milieu et techniques*, *op. cit.*, pp. 83–119. W. Ridgeway, *The Origin and Influence of the Thoroughbred Horse*, Cambridge, 1905.
32. On cattle property marks as a possible origin of writing, see A. Van Gennep, *De l'héraldisation de la marque de propriété et des origines du blason*, Paris, 1906.
33. On agriculture and the various agricultural techniques, see E. Best, *The Maori*, Wellington, 1924, 2 vols. Crozet, *Nouveau voyage à la mer du Sud...*, Paris, 1783. E. Hahn, *Die Entstehung der Pflugskultur*, Heidelberg, 1909. M. Leenhardt, *Gens de la Grande Terre*, Paris, 1937. A. Leroi-Gourhan, *Milieu et techniques*, Paris, 1945, pp. 120–37. O.T. Mason, *Woman's Share in Primitive Culture*, 1895. Ch. Robequin, *Le Than Hoa*, Paris, Bruxelles, 1929. N.J. Vavilov, 'Sur l'origine de l'agriculture mondiale d'après les recherches récentes', *Revue botanique appliquée et agriculture coloniale*, Paris, 1932. The same journal published a more elaborate account of Vavilov's ideas in 1936 under the title: 'Les bases botaniques et géographiques de la sélection', translated by A. Haudricourt.
34. B. Malinowski, *Argonauts of [...]*, op. cit.; *Coral Gardens*. On production surplus, see also M.J. Herskovitz, *The Economic Life of Primitive Peoples*, op. cit.
35. See G. Muraz, 'Les cache-sexes du Centre Africain', *Journal de la Société des africanistes*, 1932, pp. 103–12.
36. R. d'Harcourt, *Les Tissus indiens du vieux Pérou*, Paris, 1924. L. Hooper, *Hand-loom weaving, plain and ornamental*, London, 1910. F. Ikle, *Primare Textile Technicken*, Zurich, 1935. A. Leroi-Gourhan, *L'Homme et la matière*, Paris, 1943, pp. 290–309. H. Ling Roth, 'Studies in Primitive looms', *Journal of the Royal Anthropological Institute*, 1916–1918.
37. See A. Leroi-Gourhan, *Milieu et techniques*, op. cit., pp. 254–320; see also the bibliography on architecture in the present work, *Manuel d'ethnographie*, Petite Bibliothèque Payot, Paris 2002 [note 1, p. 156].

38. On the Bedouin tent, see A. de Boucheman, *Matériaux de la vie bédouine*, Damas, 1935.
39. See P. Schebesta, *Les Pygmées*, French translation, Paris, 1940.
40. On the igloo, see F. Boas, *The Central Eskimo*, Bureau of Ethnology, 6th Annual Report, Washington, 1888.
41. On the Betsileo house, a 'perpetual and universal calendar', see H. Dubois, *Monographie des Betsileo*, Paris, 1938.
42. A.C. Haddon and J. Hornell, *Canoes of Oceania*, Honolulu, 1936–1938, 3 vols. H.S. Harrison, *A Handbook to the Cases Illustrating Simple Means of Travel and Transport by Land and Water*, London, 1925. G. La Roerie and J. Vivielle, *Navires et marins. De la rame à l'hélice*, Paris, 1930. A. Leroi-Gourhan, *L'Homme et la matière*, op. cit., pp. 119–65. O.T. Mason, *Primitive Travel and Transportation*, Washington, 1896. N.W. Thomas, 'Australian Canoes and Rafts', *Journal of the Anthropological Institute*, 1905, XXXV, pp. 56–79.
43. A. Leroi-Gourhan, *La Civilisation du renne*, Paris, 1936.
44. C. Wissler, 'The Influence of the Horse in the Development of Plains Culture', *American Anthropologist*, n. s., XVI, 1914, pp. 1–25.
45. On the boat's eye, see J. Hornell, 'Survivals of the Oculi in Modern Boats', *Journal of the Royal Anthropological Institute*, 1923, LIII, pp. 289–321.

Text 11

Conceptions Which Have Preceded the Notion of Matter (1939)

M. Mauss 1939, 'Conceptions qui ont précédé la notion de matière', in Fondation pour la Science – Centre international de synthèse (eds), *Qu'est-ce que la matière? Histoire du concept et conception actuelle*, XIe Semaine internationale de synthèse. Paris: Presses Universitaires de France, 1945, pp. 15–24.

Translated by Dominique Lussier

This text represents another contribution by Mauss to those interdisciplinary 'Semaines internationales de synthèse' organised periodically by the historian Henri Berr. In dealing with the somewhat abstract topic of 'matter', Mauss built on the philosophical background he had acquired in his youth, and strengthened it with ideas and insights from classical antiquity. In a more original manner, he also drew on broader comparative perspectives on civilisations, and indeed made reference to the mythologies and cosmologies of 'so-called primitives' – which, as he noted, serve to defamiliarise us, broaden our horizon and relativise our conceptions.

By way of introduction I shall make some remarks of a philosophical order, for philosophy leads to everything, provided we can go beyond it. Philosophies and sciences are languages, and it is merely a matter of making use of the best language available. Language itself and the categories of thought are 'extractions' from the modes of thought and feeling of a given social milieu. Insofar as I believe myself to be thinking correctly, I see myself in line with the great tradition of the Encyclopedists, themselves followers of the English

School. Moreover, if our way of thinking derives at every moment from all that constitutes social life, we ought not to keep apart from it the scientific mentalité, so intimately it is connected to the mentalité as a whole. For my part I have also had the opportunity, alas, to observe the impact of social cataclysms on the progress of a science. During the First World War, I witnessed the loss of very dear collaborators and the subsequent slowdown in the forward development of the nascent sociology.

The study of a body of thought, such as that leading to the notion of matter, is purely historical. The prevailing conception of this notion, ours as well as that of the person in the street, has nothing in common with former explanations. For a Jacob Boehme, for a Cardan, lead deposits were still possessed of real life, and this was a belief they shared with the miners of their days. Nowadays, similar ideas can be found in Malaysia and among some tribes who have no difficulty whatever in conceiving that a seam might 'escape'. We can see how everything is at once continuous and discontinuous, full of contingencies and hard to predict. Each and every notion is in a perpetual state of becoming, language is continually evolving, and it is therefore highly instructive to study across various languages the history of the word that signifies 'matter'.

The word *matière* [French for 'matter'] comes from *materia*, a feminine noun in both Latin and French, while *materies* (derived, just as *materia* is, from *mater*, the feminine generative power) has yielded the word *matériau* [French for 'material']. In effect, *materies* used to be part of the vocabulary of the woodcutter and of the carpenter. It referred to the core of the tree; it was wood, the essence of all things, a solid expression connoting life. Also, it is the thread forming the woof of fabric, and it is equally the veined marble block out of which the sculptor, following the grain, will create his statue. The German word *Stoff*, a masculine noun, derives from the feminine *Estoffe* (*étoffe* in French) which also gave *Stuffe*. The Greek word *ulê* is in its origin identical to *materies*. It too is an expression used by craftsmen and artists: upon the work of the soul and of skill, a thing is extracted out of another thing. Therefore *ulê* has been translated as *materia* or as *materies*. In a recent work, I have discussed the inherent difficulties of Graeco-Latin traditions. We must bear in mind that Greek philosophy came to the Romans through commercial travellers; thus the Magus has made Aristotle say more about matter than he had originally said.

The opposition between the raw matter of the mechanic and the living matter of the craftsman-artist is only a recent one. Technical schools have recently been opened in France, and their curriculum includes the knowledge of materials. The *Technische Hochschule* in Berlin, created after the war, and other similar institutions testify to the growing interest in the study of techniques. According to Maurice Halbwachs*, man is an animal who thinks with his fingers. The work of Pierre Janet contains equally captivating formulations on this topic. Attempts have been made to link *ulê* with the current term 'technique'. But while iron forging techniques are very ancient, alloys and casting have appeared much later on, and outside of the European world (despite the prevailing opinion). Iron has been known since 2500 BC, and there are fine Han Yang bronzes dating as far back as 2200. Here again we have to upturn received notions.

We owe the connection between *ulê* and *silva* to Sir James Frazer and Lucien Lévy-Bruhl. The word is also used as a generic term for the various species of tree. But the connection is highly interesting. Silva is the germinative power conceived of as feminine: it is the forest. There lies in this notion of forest, as in that of matter, something unruly, savage, dangerous, but also something animated and receptive. This notion includes the idea of an obstacle; the forest is that which can be overcome. Prior to the discovery of metals, only fire made it possible to gain ground from the forest. Finally, as Henri Hubert remarked, the word forest implies the notion of sustained use. Not only did wood precede metal, but in our lands it was for a long time an obstacle to the spread of metallurgy, since a good wooden axe was deemed better than a metal one.

The notion of *ulê silva* is a fundamental one. Our ancestors conceived of two worlds, as still do today the Polynesians and certain tribes of North America. One world is sheltered, it is the camp – in the other world, on the outside, danger looms. On the one side there is the *ulê* and on the other, *kosmos*. The book Robert Hertz did not have time to publish, of which I possess a draft which I hope one day to publish, is an admirable study, regarding the notion of expiation, of the distinction between the world of spirits and the world of matter.

What we have best perceived so far is the dyad *matter-form*. The opposition *matter-spirit* is far more recent; it only came about with the purely mechanical and geometrical notion of matter dating perhaps from Galileo, at any rate from Descartes. It is mainly in Spinoza, in my opinion, that appears the opposition 'thought-extension'. This notion of matter purified of all spiritual elements has been developed mostly in France and Great Britain. The work of Mr Léon Brunschvicg sheds light admirably on the revolution that took place between the old and the new conception of matter.

But you are expecting me to provide you with facts borrowed from primitives, these peoples so wrongly named, that we put all on the same plane, deeming them transparent, easy to understand, lacking high intellect, illogical, mystical or else brutal or materialistic. There are among them – you can rest assured – as much divergence, if not more, than we find among ourselves, and every grade of abilities. Yet we feel profoundly disorientated when dealing with them: they teach us to think otherwise than as *homo sorbonnensis* or *oxoniensis*, thanks to the work of other *homines sorbonnenses*.

I have been working for some time on the notion of food [*nourriture*]. It is one of the originalities of French sociologists that they have studied the diverse categories of thought. Thus, food connotes subsistence (subsistance in French is used as a synonym for food), and also substance or matter. For the Romans the notion of subsistence formed the root of substance. We still find *substantia*, substance, in French legal language. The evolution of the notion is analogous in Sanskrit. The meaning of the word has changed along with the changing society. You will know Antoine Meillet's remarkable work on the correlation between these two respective evolutions. To my mind food is one of the most important notions among those underlying the notion of matter. Gorce, who is the expert, is in agreement with this view.

The fine works of Baldwin Spencer and F.J. Gillen on the Arunta of Australia, as well as those of contemporary ethnographers, allow us to extend our theory of the relationship between the notions of matter and of food. In effect, the same complex rites of initiation relating to food can be found among most primitive peoples today. Towards the age of twenty-one, the time when some of us are preparing for the 'aggregation' examinations – one of the initiation rites of the 'civilised' – the so-called savage for his part is granted the *power to eat*. In other words, he receives his initiation into food, or rather, it is necessary that the owner of a totem *opens his mouth* for him. He will then be able to eat the sacred animal when it will be revealed to him, through images or masks, what the totem is. These images and masks (depending on the tribe) thus grant a power, a new breath in relation to the animal over which they exert power. At the root of matter, as with *ulê* we find the same notions of the specificity and generality attached to food. Ideas concerning a divided power and a wandering power have always haunted the imagination, as do their offspring the ideas of error or sin. One of my pupils, undertaking research in Sainte-Thérèse, will soon be able to publish his work on the Eskimo, who make a distinction between summer food and winter food. In short, the notion of food varies according to times and places, and the notion of substance always underlies it.

There are some remarkable texts concerning the difference between the eater and the eaten. In particular, the eight volumes by Robert Hertz include Maori texts on the notion of *polkro*, for example concerning the gods who were compelled to flee 'because they have become *polkro*'. We may note that *pâpâ*, the earth, also means female. Thus *Trou-Ta-Na-A-Ragla* is the son of the earth (*pâpâ*). He has committed incest by recreating the caves together with his mother. The question is asked: how can son and mother be separated? An attempt was made by the god of the winds Ta-Cohi-Ma-Ta (or the male principle Rongo-Ma-Ta-Na). The four other gods are cowards; they would rather live and be eaten than not live at all, and therefore they now live in the form of birds and ferns. Rongo-Ma-Ta-Na is the god of agriculture, Tangaeli the god of fish; he perpetuates himself, just like the fish that people eat. The role of blood in the mythical and ritual life of the Maori is crucial. Anecdotes abound on this subject, for instance that of the English colonist told by an old Maori: 'We are somewhat related; my ancestors ate yours.'

It is an error to take myths one at a time, severing them from what has preceded them and from the forms which in their turn engender. They constitute a totality in relation to the whole collectivity. A myth is but a thread in a 'spider's web', not an entry in a dictionary. It is a matter of seeing and interpreting the ensemble. To follow the German expression, all the *Bilder* are part and parcel of the same *Bildung*. I propose to publish soon a chart of Maori mythology including the whole cosmogony, the whole cosmology, as well as a classification of the spaces and the times. Above everything, represented by circle, is found the supreme god Hio. Lower down is the part of the sky inhabited by the gods of rain and fine weather: this is the sky of the lakes. Next we descend to the place where the human spirit is fashioned. This is followed by the abode of servants of inferior gods, and then by that of inferior beings. In this way we descend to the place inhabited by souls that are ready to emerge. We finally reach the lowest place, where the spirits live side by side with things.

From this complex mythology of spaces, the following insight is also worth recalling for our purpose. There are agricultural spaces, beneath which the gods of evil kill two other gods who have not sufficiently fulfilled their role as eaters. Above, there is Koré, the mythical embodiment of all space, of which we should like to know whether it has blown on the world in order to create it, or on the contrary whether it is the world that has blown on it (the Magaïa version of the Koré myth).

There are two words to designate water in Sanskrit. One has given us *aqua*, and it was used for referring to water as a thing, the living water, female water. The other yielded *Wasser*, *water*, water that is lifeless. In all these mythologies, the division between male and female is noteworthy. The myth of Isis and Osiris is instructive in this regard (this myth still appears in Plutarch). We can say that Plato and Socrates drew inspiration at large from everything, and that on the other hand, Alexandrian thought already represents a decline. On this subject I refer you to the interesting work of Messrs Bidez and René Berthelot on Greek alchemy.

To summarise, the notion of matter arises as an animate principle, and contrary to common belief, it is according to Aristotle a living body. These forms are quite lacking in precision, but really no more so than our own conceptions of matter.

7. 'Cutting out the opossum'. P.104.ACH1. New South Wales, Australia. Mounted Haddon Collection (CUMAA©). '[The Australian] manages to catch the possum or phalanger at the top of its tree, even though the animal puts up a remarkable resistance' (Mauss, text 9).

Text 12

Techniques and Technology (1941/1948)

M. Mauss 1948, 'Les techniques et la technologie', *Journal de psychologie* 41:71–78.

Translated by J.R. Redding

Just like Mauss himself, also the psychologist Ignace Meyerson was forced by the collaborationist laws of Vichy to resign from his academic functions. Meyerson took refuge in Toulouse and organised there in 1941 a colloquium on the psychology and history of work and of techniques, with contributions from L. Febvre, A. Lalande, G. Friedmann and others. Travel restrictions and ill health prevented Mauss from attending in person, and he sent this essay instead. Mauss seems to have sensed that this was to be his last intellectual statement: besides rounding up and updating some of his long-held ideas on techniques and technology, he also touched on the aims and prospects of these studies. Both his initial comments on 'psycho-technics' and his final, almost utopic, words on planning and coordination reflect his enduring faith in humanity and in human creations, even in these darkest hours.

In order to talk meaningfully about techniques, it is first necessary to know what they are. Now there actually exists a science dealing with techniques, a science which in France does not have the place it deserves: it is the science called technology. This is worth stressing here, especially since it is the Société d'études psychologiques that is organising today's seminar on psychology and history. In these matters of psychology properly speaking, France is in fact ahead of other countries. Those of my generation have witnessed the invention – by Binet, Simon, Victor Henri, joined by Pieron then Meyerson and Lahy, and effectively continued by others – of the applications of psychology to techniques and more particularly to the recruitment of workers and technicians.

It was only after the war of 1914 that 'psycho-techniques' – perfected in America and developed everywhere – returned to blossom in France, especially in Paris, where important initiatives secured considerable and indeed essential results. It is clear that the psychology of techniques currently carried out corresponds to a specific moment in the history and the nature of these techniques. Now while this aspect of the study of techniques is of French origin, the same cannot be said of the overall development of the science of which it is but an element, namely technology. It is clear that the psychology of techniques that is currently undertaken is but a moment in the history and the nature of those techniques.

Technology is a science much more developed in other countries. It rightly claims to study all techniques, the entire technical life of humankind since human origins and down to our present day. Technology is both at the base and the apex of all research undertaken on this object. 'Psycho-technique' is thus simply a technique of techniques. It nevertheless presupposes a thorough overall knowledge of the general subject, that is of techniques.

It is necessary above all to specify the place of technology, the work it has produced, the results already obtained, and the extent to which it is essential for any study of humankind – including studies of the human psyche and of human societies with their economy, their history, the land from which they derive their livelihood, and, in consequence, with their mentality. It is not because technology is not regularly taught in France that we should abstain from discussing it here (I know of only one teaching course, but it is very elementary and, moreover, geared at observing the techniques of so-called primitive or exotic people).

This science of technology was in reality established in Germany. As the country of predilection for the historical and scientific study of techniques, Germany remains, along now with America, at the forefront of all technical progress. In truth, this science was instituted by Franz Reuleaux, the great theorist and mathematician, engineer and technician of mechanics. The Prussian authorities responded immediately to his appeals, and under his direction was opened the first of the Advanced technical schools (*Technische Hochschulen*) – that of Berlin, with university rank and a diploma (Dipl. Ing.) equivalent to the doctorate. The general study of technology, its theory and history, is mandatory there in all the special sections leading to the different diplomas. Therein lies the natural basis for the general study of techniques, and we would do well to recognise this here.

In our country, however, even in our most worthy scientific establishments, even in our illustrious and still glorious Conservatoire des Arts et Métiers, technology does not have the position of a general theory of crafts [*métiers*]. At the Musée des antiquités nationales at Saint-Germain-en-Laye, my much-lamented brother-worker Henri Hubert had installed the 'Salle de Mars', dedicated to the comparative art and ethnology of the Stone Age: at the present moment, this room is no longer even in use. At the Musée de l'Homme we have succeeded with the help of the Institut d'Ethnologie in making considerable improvements, but even this remains modest. The Vienna Museum, the Pitt Rivers Museum, Nordenskiöld's Museum in Gothenburg are in many respects better placed than we are.

As for the theory or the historical, geographic, economic and political descriptions of crafts, various attempts have been made in France to launch it, but without success. We have not even kept up the tradition of those good popular histories of industry as recounted by the likes of Becquerel and Louis Figuier, which, however anecdotal, were instructive to the young and even to children. My uncle Durkheim made me read them. One of those on the right tracks, my former teacher Alfred Espinas, gave a course on these matters in Bordeaux which I well remember – his book on *Les origines de la technologie* is still worth reading. He did not, however, develop his ideas enough nor pursue his research widely and deeply enough.

* * *

A few remarks will indicate the paths already opened, and the directions to which they lead. Let us suppose that there is a great number of known facts which several among us might perhaps not know. At a time when techniques and technicians are fashionable – in contrast to so-called pure science and philosophy, accused of being dialectic and sterile – it would be necessary, before extolling the technical mind, to know what it is.

To begin with, here is a definition: *We call 'technique' an ensemble of movements or actions, in general and for the most part manual, which are organised and traditional, and which work together towards the achievement of a goal known to be physical or chemical or organic.* This definition aims to exclude from consideration those religious or artistic techniques, whose actions are also often traditional and even technical, but whose aim is always different from a purely material one and whose means, even when they overlap with a technique, always differ from it. For example, rituals of fire can control the technique of fire.

Looking at techniques in this way allows us to classify them and to provide a comparative table of what we still call *labours, arts* and *crafts* [*traveaux, arts et métiers*]. Thus, we refer to the craft of the painter, even when speaking of a painter of pure art. This definition enables us to classify the different parts of technology.

In the first place we have descriptive technology. This relates to:

1. Sources that are classified historically and geographically, such as tools, instruments, machines; the last two being analysed and assembled.

2. Sources that are studied from physiological and psychological points of view, including the ways in which they are used, photographs, analyses, etc.
3. Sources that are organised according to the system of industry of each society studied, such as food, hunting, fishing, cooking, preserving, or clothing, or transport, and including general and specific usages, etc.

Onto this preliminary study of the material of techniques, must be superimposed the study of the function of these techniques, their interrelations, their proportions, their place in social life.

These last-named studies lead to others. They allow us to determine the nature, the proportions, the variations, the use and the effect of each industry, its values within the social system. All these specific analyses really do guide us towards more general considerations. They make it possible to propose different classifications of industries and, even more importantly, they enable us to classify societies in terms of their industries.

From this follows a third order of general considerations. A growing number of scholars (ethnologists, anthropologists, sociologists, etc.) attach high importance to comparisons between societies with similar industries. They believe they can prove the borrowings of such or such industries, the *areas of distribution* of others, and even the *historical layers of distribution*, as prehistorians have already proposed. Some cautious and even very cautious scholars like the Americans, record the facts and occasionally make historical deductions from them. Others, less prudent, have sought to reconstruct the entire history of humanity through the history of techniques. They have gone so far as to speak of a Stone Age in the Congo belonging to the period of civilisation when inheritance rights descended through the female line.

These excesses, however, do not detract from the soundness of the method when properly carried out. Even with regard to the most primitive societies known to us, techniques, their functions propagated and then preserved by tradition, are – since Boucher de Perthes – the best method of classifying societies, even chronologically. The *Sinanthropos*, Peking Man, could cook with fire, a sure proof that this being was human. We do not know if he could speak but this is probable given that he could maintain some way of keeping the fire going.

I myself have put forward some views on the techniques of the body and their functions. For example, techniques of swimming are variable, and allow us to classify whole civilisations. All techniques are specific to each civilisation, as both the tools and their handling are infinitely variable. Thus, at the same time as they are human by nature, techniques are also characteristic of each social condition.

I know that some see a mystery in all this. Granted the notion of *homo faber*. But Henri Bergson's idea of creation is the exact opposite of that of technicity, of creation starting from matter which man has not created, but which he adapts and transforms and which is assimilated by the common effort – an effort constantly and everywhere fed by new contributions. From this established point of view, the expression '*Ars homo additus naturae*' is even more true of arts and crafts than it is of art. It is through the penetration of physical nature that art and craft come about, that the artisan and industrialist gain their livelihood, and that come about the development of industry and of civilisations, indeed of civilisation itself.

* * *

There is another perspective from which the study of techniques – technology – is even more important. It is that of its relations with the sciences, the daughters and mothers of techniques. In fact, today, a vast majority of humankind is increasingly engaged in such occupations. The greater part of its time is caught up in this work, whose treasure of tradition is safeguarded and increased by society. Even science, especially the wonderful science of today, has become a necessary element for techniques, a means. We can 'hear' or 'see' electrons or ions by virtue of a technique incorporated in every radio. A precision engineer nowadays operates lines of sight and does vernier readings which had before been the prerogative of astronomers. An aeroplane pilot reads maps of a new kind, and at the same time sees the tops of mountains or into the depths of the sea in ways which none of us could have dreamed of in our young days. The nineteenth-century hymn to science and craftsmanship is even more true for the twentieth. The exhilaration of production has not vanished. There are machines which are truly beautiful and good, and fine-looking cars. There are excellent crafts done with machines. There is pleasure in the work, in the accurate calculation, in the perfection of the manufacture and its mass production, using machines invented according to precise plans, from accurate blueprints – machines used to produce in series even more precise and immense or compact machines, which themselves serve to manufacture others, in a never-ending chain in which each is but a link. That is what we are living through now, and this is not over yet.

If we add that nowadays even the most elementary technique, such as that relating to food production (about which we know something) is becoming integrated in these great cogwheels of industrial plans, if we note that the 'industrial economy', the one still wrongly considered a part of so-called political economy, is turning into an essential gearing in the life of every society, and even of the relations between societies (ersatz, etc.), we can begin to measure the vast contribution of technique to the development of the mind [*esprit*].

Thus, since that very remote period of time when the *Sinanthropos* cave dweller from Chou-Kou-Tien, near Peking (the least human of the human beings known to us) could at least keep fire going, the sure sign of humanity has been the existence of techniques and their traditional perpetuation. There exists indeed a definitive classification of humans, and it is the classification of their techniques, their machines, their industries, their inventions. In this progress resides the spirit, the science, the strength, the skills, the greatness of their civilisation.

Let us neither praise nor blame. There is more to collective life than techniques, but the predominance of such or such technique at a given stage of humanity is a feature that qualifies nations. In a fine piece of work published in the *Revue de naturalistes* one of our good 'comparativists', André-Georges Haudricourt, has demonstrated how the best harnessing techniques for oxen or horses have all spread slowly out of Asia. In this Asia has always been superior, and in many other matters it still remains a model.

We can even address these questions in quantitative terms. The number of patents applied for and granted in France which are being recognised elsewhere is, alas, much smaller than the numbers of German, English and especially American patents. It is the latter who lead the pack and set the pace. Even science itself is becoming more and more technical, and increasingly influenced by techniques. The purest strands of research have immediate applications. Everyone knows about radioactivity. We are now able to conserve and concentrate neutrons, and soon perhaps we shall see their exploitation. Electronic microscopes can now magnify by the power of a million, and it will soon be possible to photograph atoms. We see and we test with them. The circle of science-technical relations is growing ever wider, but at the same time also tighter.

It only remains to master the unleashed demon. The danger may, however, be exaggerated. Let us talk neither of good or evil, morality, law, force, money, currency reserves or stock exchanges. All this is less important than what the future holds in store for us. At the present time, the future belongs to research and development departments of the kind found in big companies. These research departments must have the closest possible ties with statistical and economic institutions, for no industry can exist other than in relation with many others, with numerous sciences and with strong regulated economies, whether individual or public. The various plans of action now in existence are more than a fad, they are necessities. Techniques are already independent, better still; they are in a category by themselves with a place of their own, no longer merely swayed by happy accidents or fortuitous hazards of interests and inventions. Techniques take their part in pre-planned projects where huge buildings house immense machines for manufacturing others, which in turn manufacture further power or precision machines, each dependent on the other and destined to make products under specifications as constraining as the products of the laboratories of old.

But the coordination of the ensemble of these projects cannot be left to chance. Techniques intermingle, with the economic base, the workforce, those parts of nature which societies have appropriated, the rights of each and of everyone – all crosscut each other. From now on, as has already happened in some countries, will rise above individual projects the silhouette of the grand plan, of so-called plannification. I can still see an inspired François Simiand*, then assistant to Albert Thomas at the Ministry of Armaments in the last war, calculating worldwide statistics as well as the military or civil needs of the country, deciding what was possible and what was useless. This has been rightly called a wartime economy, but the methods launched then have progressed, not only in wartime when they are necessary, but also in peacetime.

And whoever talks of a plan is talking about the activity of a people, a nation, a civilisation, and speaking, better than ever, of morality, truth, efficiency, usefulness, the good. It would be futile to oppose matter and mind, industry and ideals. In our times, the power of the instrument is the power of the mind, and its use implies morality as well as intelligence.

8. 'Drinking'. P.17.ACH1. New South Wales, Australia. Mounted Haddon Collection (CUMAA©). 'The constant adaptation to a physical, mechanical or chemical aim (e.g. when we drink) is pursued in a series of assembled actions, and assembled for the individual not by himself alone but by all his education, by the whole society to which he belongs, in the place he occupies in it' (Mauss, text 9).

Biographical Notes

The following pages provide succinct biographical and bibliographical information on Marcel Mauss himself, as well as on the various authors he invoked as sources, references or protagonists (and whose first mention in this volume is followed by an asterisk, as explained in the Preface). While these authors are listed here in a straightforward alphabetical order, most of them belong to three relatively coherent research traditions: French, German and American (others being British or Scandinavian). The following references provide some background information and historical studies on these traditions. So far as French anthropology and social sciences are concerned, aspects of their development are discussed by Mucchielli (1998), Clifford (1988) and Lebovics (1992), as well as in the memoirs of Haudricourt (Haudricourt and Dibie 1987) and of Leroi-Gourhan (1982, 2004, and Audouze and Schlanger 2004). See also the references provided in the entries for Mauss and for Durkheim. The history of anthropology in the Anglo-Saxon world has been researched by, among others, Kuper (1973), Stocking (1987) and Trigger (1989), and see Hinsley (1981) and Schlanger (1999) for early North American Anthropology and the Bureau of American Ethnology. Relevant aspects of German social sciences are discussed by Herf (1984), Smith (1991) and Zimmerman (2001).

Balfour, Henry (1863–1939)

As the longstanding curator of the Pitt Rivers Museum in Oxford, Balfour pursued and amplified the anthropological interests of his predecessors General A.L. Pitt Rivers and E.B. Tylor. Through the judicious seriation of ethnographic and archaeological specimens, he attempted to display both the geographical spread and the logical evolution of technical objects and material culture – maintaining a museum whose coherence and coverage Mauss admired. In this vein, Balfour published on *The Evolution of Decorative Arts* (1893) and *The Natural History of the Musical Bow* (1899).

Bastian, Adolf (1826–1905)

Trained as a Navy doctor, Bastian went on to create the German *Völkerkunde* School. Besides contributing to its institutionalisation (university chairs, museum, learned societies and publications), he also gave the discipline its foundations in both geography and philosophy. Emphasising the study of objects and languages, Bastian promoted the notion of *elementargedanken* as primordial, universal ideas which then become differentiated according to the circumstances of each *geographische provinzen*, and following processes of migration and borrowing.

Bergson, Henri (1859–1941)

From the Ecole Normale Supérieure and later the Collège de France, Bergson articulated a compelling philosophy of *élan vital* and *énergie spirituelle* which gained remarkable popularity with the publication of *Evolution créatrice* (1907) and *Les deux sources de la morale et de la religion* (1932). This intuitionism and anti-rationalism met with strong opposition from Durkheim and his circle. On matters of substance, Mauss specifically crossed swords with Bergson over his famous concept of *homo faber*, criticised for its individualistic and Nietzschean undertones.

Boas, Franz (1858–1942)

Trained in natural sciences in Germany, Boas emigrated early in his career to the USA and eventually established there his 'four fields' conception of anthropology. From Columbia University and the American Museum of Natural History in New York, he opposed the universalist evolutionism of L.H. Morgan and J. W. Powell and advocated instead a particularistic and historicist view of cultures, conceived in the plural and without value judgement. The methodological implications of this approach, largely derived from the German *Völkerkunde*, included comparative studies of cultural traits, in their temporal and spatial distributions. Boas's ethnographic research on Pacific and Eskimo mythology, languages and arts (e.g. *The Central Eskimo*, 1888, or *The Kwakiutl of Vancouver Island*, 1909) proved influential for several of Mauss's own works on religion, on social morphology and on gift giving and exchange.

Cushing, Frank Hamilton (1857–1900)

A member of the Bureau of American Ethnology, Cushing is well known for his work among the Zuni of New Mexico. The 'participant ethnography' he pioneered there gave him insights into Zuni mythology and cosmology which served Durkheim and Mauss in their work on Primitive Classification. Himself a keen experimenter of ancient techniques and procedures (e.g. 'The Arrow', 1895), Cushing also wrote the highly original 'Manual Concepts: A Study of the Influence of Hand-usage on Culture Growth' (1892).

Durkheim, Emile (1858–1917)

A founder of modern sociology, Durkheim was trained in philosophy at the Ecole Normale Supérieure (alongside Jaurès, Bergson, Janet and others). Based in Bordeaux from 1887 to 1902, he wrote there his thesis on *De la division du travail social* (1893), as well as his *Règles de la méthode sociologique* (1895), followed by *Le suicide* (1897). As part of his ambition to establish a comprehensive and scientific sociology, Durkheim launched in 1898 the yearly *Année sociologique*, an extraordinary collaborative effort and intellectual achievement. From the end of the 1890s onwards, and following his return to Paris and the Sorbonne in 1902, Durkheim began to emphasise religion as the primary factor in social life. This conception culminated with *Les formes élémentaires de la vie religieuse* (1912), which drew extensively on Australian ethnography. The death of his son in the battlefields of the First World War (like that of several other young collaborators) exacerbated his chronic ill health and precipitated his death in 1917.

For more information on Durkheim's life and works, see principally Lukes (1973 and subsequent editions), as well as Fenton (1984), Pearce (1989), Pickering (1994, 2003). Durkheim's correspondence with Mauss has recently been published (Durkheim 1998). On the *Année sociologique* and the Durkheimians before and after the First World War see Besnard (1983), Parkin (1996), Marcel (2001).

Espinas, Victor Alfred (1844–1922)

Philosopher and sociologist, Espinas was an early promoter of the social sciences in France, notably as Durkheim's senior colleague and Mauss's teacher at Bordeaux. His publications include *Des sociétés animales: étude de psychologie comparée* (1877), and particularly *Les origines de la technologie* (1897), in which the analysis of Greek classical sources led him to more philosophical concerns over the role of technical activities in the evolution of human thought. This 'praxeology', encompassing all practices and efficient actions, had however little impact on the social sciences of his times.

Febvre, Lucien (1878–1956)

Founder with Marc Bloch of the journal *Annales d'histoire économique et sociale* (1929), Febvre promoted the *Annales* conception of history as a synthesis of social, economic and mental factors. In close contact with the Durkheimians, he had attempted to clarify debates over social morphology, 'genre de vie' and human geography in *La Terre et l'évolution humaine* (1922). Besides organising a collective reflection on *Civilisation, le mot et l'idée* (1930), to which Mauss contributed, Febvre also wrote a brief but influential 'Réflexion sur l'histoire des techniques' (1935) in the *Annales*.

Frobenius, Leo (1873–1938)

The ethnographer Frobenius helped consolidate and popularise the German culture-historical school. Following numerous expeditions to Africa (notably the Congo and West Africa), he identified there a series of root cultures through the *Kulturmophologie* method, at once carefully documented through distribution maps and at the same time highly impressionistic. Like Bastian and others, he expounded a biologising view of culture, assimilated to an organism with its overall form and its life cycles. His Institute for Cultural Morphology issued lavish publications on ethnographic and archaeological African arts.

Graebner, Fritz (1877–1934)

An historian by training, Graebner was influenced by Bastian and Ratzel to formalise the culture-historical method of *Kulturkreise*, whereby cultural traits are invented once (as 'elementary ideas'), combine with others into cultural patterns, and then gradually spread from their epicentre outwards – as in the case of the Oceanian *Bogenkultur*. In his *Die Methode der Ethnologie* (1911), Graebner advocated the critical examination of historical and ethnographic sources. As with other protagonists, Mauss made use of the evidence they meticulously accumulated, but remained wary of the interpretations proposed.

Halbwachs, Maurice (1877–1945)

Trained as a philosopher, Halbwachs was considerably influenced by Bergson, as well as by Durkheim, and he later contributed to establish the *Annales* school with Febvre. He published studies on the economic aspects of social life, such as *La classe ouvrière et les niveaux de vie* (1912) and *Morphologie sociale* (1938), as well as on the phenomenon of collective memory in *Les cadres sociaux de la mémoire* (1925).

Haudricourt, André-Georges (1911–1996)

A highly original continuator of Mauss's technological approach, Haudricourt was sent by his teacher to study plant domestication and genetics under N. Vavilov in Soviet Russia. An atypical Marxist, linguist, agronomist and orientalist, as well as historian of science and of techniques, Haudricourt wrote on such topics as 'De l'origine de l'attelage moderne' (1936), 'Relations entre gestes habituels, formes des vêtements et manière de porter les charges' (1948), and 'La technologie, science humaine' (1964).

Hertz, Robert (1881–1915)

Trained in philosophy at the Ecole Normale Supérieure, Hertz joined the Durkheimians and the *Année sociologique* group, where he specialised in religious sociology. He wrote original and influential papers on 'Contribution à une étude sur la représentation collective de la mort' (1907) and 'La prééminance de la main droite. Etude sur la polarité religieuse' (1909), and he also begun to study religious folklore and folktales before his tragic death on the battlefields of the Somme. Mauss, a great admirer, brought to posthumous publication parts of Hertz's thesis on *Le péché et l'expiation dans les sociétés primitives* (1922).

Holmes, William Henry (1846–1933)

With wide-ranging interests in geology and biology, and considerable artistic talent, Holmes joined the Bureau of American Ethnology early on and became its director upon Powell's death. Besides his publications on *Pottery of the*

Ancient Pueblos (1886) and *Handbook of Aboriginal American Antiquities* (1919). Holmes also drew on biological and ethno-archaeological perspectives to inaugurate the dynamic study of lithic technology, which enable him to argue for a recent (post-Palaeolithic) date for human antiquity in America.

Hubert, Henri (1872–1927)

An archaeologist and sociologist of religion, Hubert was an important contributor to the *Année sociologique* (notably to the rubric 'technology'), and he was also Mauss's closest collaborator. In their *Mélanges d'histoire des religions* (1909), Hubert and Mauss assembled several early texts on sacrifice and on magic (and their relations with techniques). As curator at the Musée des Antiquités Nationales in Saint-Germain-en-Laye he set up exhibition halls dedicated to the comparative studies of objects, techniques and cultures. He also lectured on the prehistoric ethnography of Europe, and published posthumously on *Les Celtes et l'expansion celtique à l'époque de la Têne* (1932).

Leroi-Gourhan, André (1911–1986)

Following his initial training in palaeontology and oriental studies, Leroi-Gourhan became a student of Mauss and specialised in ethnographic studies of techniques around the Pacific. In 1936 he contributed to the *Encyclopédie française* a brief essay on techniques, classified not according to their goals but rather their 'means of action on matter' – an approach expanded in the two volumes *Evolution et techniques* (1943, 1945). From the Musée de l'Homme, the Sorbonne and later the Collège de France, and through such works as *Le Geste et la parole* (1964, 1965), Leroi-Gourhan contributed to establish the French school of technology in anthropology and in archaeology.

Lévy-Bruhl, Lucien (1857–1939)

As a philosopher and psychologist, Lévy-Bruhl propounded a theory which differentiated 'primitive' and 'civilised' or 'logical' modes of thought (e.g. *La mentalité primitive*, 1922). This distinction gave rise to considerable opposition from various quarters, and Mauss's discussion of the rationality behind technical knowledge can be seen at last in part as a reaction. This disagreement aside, both scholars collaborated (together with Paul Rivet) in the creation in 1925 of the then influential Institut d'Ethnologie of the University of Paris, where Mauss did most of his teaching.

Mason, Otis Tufton (1838–1908)

A member of Powell's team, Mason was for many years the curator of the ethnographic collections at the Smithsonian Institution. Besides being implicated in a defining debate with Franz Boas over museum arrangements and the interpretation of ethnographic variability, he authored important technological studies and compendia such as *Aboriginal American Basketry: Studies in a Textile Art without Machinery* (1902).

Mauss, Marcel (1872–1950)

Mauss was initiated into sociology and anthropology by his uncle and mentor, Emile Durkheim. From Bordeaux and later Paris, he collaborated in the creation of the *Année sociologique*, and specialised initially in studies of 'primitive' religions, on which he lectured at the Ecole Pratique des Hautes Etudes and published with H. Hubert. Back from active duty in the First World War, Mauss took the task of maintaining and expanding the intellectual and institutional achievements of the school upon Durkheim's death. He co-established in 1925 the Institut d'Ethnologie, and was elected in 1931 to the Collège de France. While retaining some core Durkheimian principles, Mauss broadened the sociological domain towards psychology, and particularly towards social anthropology and ethnology, of which he is a recognised founder. His best known work is probably the 'Essay on the Gift' (1925a); Mauss's singling out of the 'total prestation' as the social, moral and religious foundations of human existence has been crucial to the development of Lévi-Straussian structuralism, as well as recent Marxist, economic and gender approaches. Mauss has also published (see the Bibliography) important and influential essays on Primitive Classification (1903, with E. Durkheim), on Eskimo Seasonal Variations (1906), on The Relations Between Psychology and Sociology (1924), on the Notion of the Person (1938), and on the Techniques of the Body (1935) – here reproduced alongside Mauss's other statements on techniques, technology and civilisation. Following the German Occupation of Paris, Mauss was forced to retire from his academic functions: he died in 1950 after a long and debilitating illness.

Besides his own biographical presentation (Mauss 1930/1998) major publications on Mauss's life and works include first and foremost the richly documented biography by Fournier (1994), as well as the slimmer volume by Tarot (2003) and the more philosophically oriented analyses of Karsenti (1994, 1997). In English, Gane (1992) provides an important perspective on the 'radical' aspects of Mauss's work. The volume edited by James and Allen (1998) covers various aspects of Mauss's contribution and legacy, notably in the Anglo-Saxon world. Carrithers et al. (1985) have discussed the ramification of the notion of the person, while Allen (2000) has recently assembled a series of analysis and commentaries on Mauss's works on categories. Most of Mauss's fairly dispersed writings have been collected together and edited by Karady in three useful volumes (Mauss 1968, 1969), which complement the collection of *Anthropologie et sociologie* (1950) and the *Manuel d'ethnographie* (1947). More recently, the ensemble of Mauss's political writings (in newspapers and pamphlets) has been assembled by Fournier (Mauss 1997). Some previously unpublished texts written by Mauss for the new series of the *Année sociologique* (1924) have recently been published by J. Mergy in the *Année sociologique* (2004). To the previous translations of Mauss's major texts into English have recently been added the publication by the Durkheim Press of *On Prayer* (2003), *The Nature of Sociology* (2005), the *Manual of Ethnography* (2006) and, of course, the texts assembled in the present volume.

Meyerson, Ignace (1888–1983)

Originally trained in physiology and natural sciences, Meyerson launched the study of comparative historical psychology, influenced in part by Durkheimian social sciences. Seeking to understand the psychological specificity of humanity, he conducted in the late 1920s several experiments on the uses of instruments by monkeys. Through his long-term association with the French society of psychology and its *Journal de psychologie normale et pathologique*, Meyerson provided Mauss with a platform for fertile interdisciplinary contacts. Soon after the Second World War he published *Les fonctions psychologiques et les oeuvres* (1947) which considerably influenced sociologists and historians of mentalities.

Montandon, Georges (1879–1944)

A Swiss medical doctor, Montandon came to ethnology through his explorations in Africa and particularly in Siberia and Japan. Teaching at the conservative Ecole d'Anthropologie, he promoted a German-inspired version of diffusionism, with cycles of differentiating cultures. His *L'Ologénèse culturelle. Traité d'ethnologie cyclo-culturelle et d'ergologie systematique* (1934) includes what was for its time the most detailed and best documented presentation of traditional techniques, instruments and weapons. Initially of communist and egalitarian leanings, Montandon veered in the 1930s towards an extreme form of racism and anti-Semitism, manifested also in his scientific writings. He actively collaborated with the Vichy implementation of Nazi policies, and was executed by the Resistance.

Nordenskiöld, Erland Nils Herbert (1877–1932)

Studies of geology and palaeontology led Nordenskiöld to South America, where he researched indigenous cultures, their origins and evolution, and came in close contact with Rivet and Alfred Métraux. Nordenskiöld's observations on borrowings and exchanges of material culture were highly regarded by Mauss, as was his innovative museology at the Göteborg ethnographic museum.

Powell, John Wesley (1834–1902)

Soldier and geologist, surveyor of the Colorado, Powell became interested in the native inhabitants of the region. In 1879 he created the Bureau of American Ethnology, in association with the Smithsonian Institution and the US Congress. Alongside his pragmatic organisational capacities, Powell developed under the influences of L. Ward and of L.H. Morgan a sweeping evolutionist conception of a unified science of humanity. As he explained in the *Annual Reports of the Bureau of American Ethnology* and in other publications, one of the most important branches of this science was 'technology', understood as the science of activities geared to the production of well-being.

Ratzel, Friedrich (1844–1904)

Upon extensive travelling across Europe and America, Ratzel dedicated himself to geographical studies in Munich and Leipzig, and published such works as *Anthropogeographie* (1882–91) and *Politische Geographie* (1897). With his focus on boundaries and Lebensraum, Ratzel contributed to the establishment of modern political geography. While developing a theory of migration to account for historical processes, he also emphasised the role of environmental factors as determinants of human activity. This was understood as an alternative to the social morphology of the Durkheimians, and Ratzel was actually invited to contribute on the subject in the *Année sociologique*.

Reuleaux, Franz (1829–1905)

An influential mechanical engineer, Reuleaux became the Rector of the pioneering Technical University of Berlin, where he helped to promote German industry and establish its patent system. Reuleaux specialised in kinematics, the analysis of the forces and motions of machines. His principal publication, *The Kinematics of Machinery, Outlines of a Theory of Machines* (1875), included an important analysis and descriptive vocabulary which influenced Mauss.

Rivers, William Halse (1864–1922)

Trained in medicine and neurology, the Cambridge-based Rivers undertook important research in physiology and experimental psychology. Following his participation in the pioneering Torres Straits expedition in 1898, he specialised in anthropological questions of kinship and social organisation in Asia and Melanesia. After the First World War, Rivers endorsed the controversial diffusionist theories of G.E. Smith and W.J. Perry on the Egyptian origins of all civilisations.

Rivet, Paul (1876–1958)

Sent as a young medical doctor to South America, Rivet jointed the staff of the Muséum d'Histoire Naturelle. From the chair of '*Ethnologie des hommes actuels et des hommes fossiles*', he transformed the Musée du Trocadero into the Musée de l'Homme (1937), an innovative cultural and museological institution (thanks notably to G.H. Rivière). Alongside his linguistic and diffusionist research on the early settlement of the American continent, Rivet pursued a political career as a socialist deputy and as the co-founder of the *Comité de vigilance des intellectuels antifascistes* (1934).

Schmidt, Wilhelm (1868–1954)

A German anthropologist specialising in the Asiatic and Oceanic world, Schmidt lived mainly in Austria and established there the Austrian variant of the *Kulturekreise* culture-historical school. A Roman Catholic priest interested in the origins of monotheism and the importance of religious phenomena as a distinguishing characteristic of cultures across wide spatial and temporal eras,

Schmidt advanced views which Mauss found particularly objectionable. He wrote *The Origins and Growth of Religion* (1931) and *The Culture-Historical Method in Ethnology* (1939).

Simiand, François (1873–1935)

A long-standing collaborator of Durkheim and the *Année sociologique*, Simiand was an economist, specialising notably in labour relations and wages. During the First World War he illustrated a form of civic engagement by becoming a high-level functionary in the Ministry of Armament, from where he contributed his statistical expertise to industrial planning and production. Simiand later taught these topics at the Conservatoire National d'Arts et Métiers.

Sorel, Georges (1847–1922)

An engineer by training, Sorel was a radical political commentator and philosopher who initially endorsed a form of Marxist historical materialism, and engaged in a dialogue with both Durkheim and Mauss. He soon veered to an extreme form of anarcho-syndicalism, with a Bergsonian-inspired attention to the irrational and instinctive élan of the masses. This led him to promote the notion of the myth as a primitive form of consciousness, and to emphasise the creative values of strikes and sabotage in his *Réflexions sur la violence* (1908).

Spengler, Oswald (1880–1936)

Philosopher and historian, Spengler gained worldwide notoriety in the aftermath of the First World War, with the publication of *The Decline of the Occident* (1918). Here and in subsequent works he expounded organicist ideas shared with Bastian and Frobenius on the birth, growth and decay of civilisations, an inevitable life-cycle which fostered the emergence of new and distinctive cultural forms. In *Der Mench und Die Technik* (1931) Spengler presented techniques in an almost Bergsonian light, as an immemorial 'vital tactic' where both instruments and activities were implicated in a struggle for the mastery of life.

9. 'Holding the snake before biting the head'. P.94.ACH1. New South Wales, Australia. Mounted Haddon Collection (CUMAA©). 'In all these elements of the art of using the human body, the facts of education were dominant' (Mauss, text 9).

10. 'An old man explaining the meaning of the sacred mark'. P.129.ACH1. New South Wales, Australia. Mounted Haddon Collection (CUMAA©). 'Technical actions, physical actions, magico religious actions are confused for the actor' (Mauss, text 9).

11. 'The kangaroo and emu were painted and cut on many trees and rocks during the ceremony'. P.136.ACH1. New South Wales, Australia. Mounted Haddon Collection (CUMAA©). 'As you will know, the Australian manages to outrun kangaroos, emus, and wild dogs' (Mauss, text 9).

General Bibliography

This bibliography lists the references indicated in the Preface, the Introduction and the Biographical Notes. The references provided in Mauss's own publications are listed there (in texts 9 and 10).

Adas, M. 1989. *Machines as the Measure of Man: Science, Technology and Ideologies of Western Dominance*. Ithaca: Cornell University Press.

Allen, N.J. 2000. *Categories and Classifications. Maussian Reflections on the Social.* Oxford: Berghahn Books.

Andrews, H.F. 1993. 'Durkheim and Social Morphology', in S. Turner (ed.), *Emile Durkheim, Sociologist and Moralist*. London: Routledge, pp. 111–35.

Antlieff, M. 1993. *Inventing Bergson. Cultural Politics and the Parisian Avant-garde.* Princeton: Princeton University Press.

Audouze, F. and N. Schlanger (eds) 2004. *Autour de l'homme: contexte et actualité d'André Leroi-Gourhan.* Nice: Editions APDCA.

Balfet, H. 1975. 'Technologie', in R. Cresswell (ed.), *Eléments d'ethnologie*, vol. 2. Paris: Armand Colin, pp. 44–79.

Bataille, G. 1929. 'Le gros orteil', *Documents* 1/6:297–302.

______ 1930. 'Le bas matérialisme et la gnose', *Documents* 2/1:1–9.

Benda, J. 1927. *La Trahison des clercs*. Paris: Grasset.

Bergson, H. 1907. *Evolution créatrice.* Paris: Presses Universitaires de France.

______ 1932. *Les Deux sources de la morale et de la religion.* Paris: Presses Universitaires de France.

Besnard, P. (ed.) 1983. *The Sociological Domain: the Durkheimians and the Founding of French Sociology*. Paris: Editions de la Maison des sciences de l'homme/Cambridge: Cambridge University Press.

Birnbaum, P. 1972. 'Du socialisme au don', *L'Arc* (Marcel Mauss) 48:41–46.

______ 1993. *La France aux Français - Histoire des haines nationales.* Paris: Seuil.

Burwick, F. and P. Douglass (eds) 1992. *The Crisis in Modernity. Bergson and the Vitalist Controversy*. Cambridge: Cambridge University Press.

Carrithers, M., S. Collins, and S. Lukes (eds) 1985. *The Category of the Person: Anthropology, Philosophy, History*. Cambridge: Cambridge University Press.

Childe, V.G. 1933. 'Is Prehistory Practical?', *Antiquity* 7:410–14.

Clifford, J. 1988. *The Predicament of Culture. Twentieth-Century Ethnography, Literature and Art*. Cambridge, MA: Harvard University Press.

Cresswell, R. 1976. 'Techniques et culture: les bases d'un programme de travail', *Techniques & Culture* 1:2–6.

______ 1996. *Prométhée ou Pandore ? Propos de technologie culturelle.* Paris: Kimé.

Cruickshank, J. 1982. *Variations on Catastrophe: Some French Responses to the Great War.* Oxford: Clarendon Press.
Durkheim, E. 1893. *De la division du travail social.* Paris: Presses Universitaires de France.
——— 1897a. *Le suicide.* Paris: Presses Universitaires de France.
——— 1897b. 'La conception matérialiste de l'histoire – une analyse critique de l'ouvrage d'Antonio Labriola, Essai sur une conception matérialiste de l'histoire', *Revue philosophique* XLIV:645–51.
——— 1901. 'Technologie', *Année sociologique* 4:593–94. **Text 1.**
——— 1912. *Les formes élémentaires de la vie religieuse.* Paris: Presses Universitaires de France.
——— 1988. *Les règles de la méthode sociologique.* Paris: Editions Flammarion, (republication of the first edition, 1895, and second edition 1901, Presses Universitaires de France).
——— 1998. *Lettres à Marcel Mauss.* Présentées par Philippe Besnard et Marcel Fournier. Paris: Presses Universitaires de France.
Durkheim, E. and M. Mauss 1903. 'De quelques formes primitives de classification. Contribution à l'étude des représentations collectives', *Année sociologique* 6:1–72.
——— 1913. 'Note sur la notion de civilisation', *Année sociologique* 12:46–50. **Text 3.**
Espinas, A. 1897. *Les origines de la technologie.* Paris: Alcan.
Febvre, L. 1922. *La Terre et l'évolution humaine. Introduction géographique à l'histoire.* Paris: Albin Michel.
Fenton, S. 1984. *Durkheim and Modern Sociology.* Cambridge: Cambridge University Press.
Fournier, M. 1994. *Marcel Mauss.* Paris: Fayard.
Fox, R. and G. Weisz (eds) 1980. *The Organization of Science and Technology in France, 1808–1914.* Paris: Editions de la Maison des sciences de l'homme/Cambridge: Cambridge University Press.
Gane, M. (ed.) 1992. *The Radical Sociology of Durkheim and Mauss.* London: Routledge.
Guillerme, J. and J. Sebestik 1968. 'Les commencements de la technologie', *Thalès* 12:1–72.
Haudricourt, A.-G. 1936. 'De l'origine de l'attelage moderne', *Annales H. E. S.* 8:515–22.
——— 1948. 'Relations entre gestes habituels, formes des vêtements et manière de porter les charges', *La Revue de géographie humaine et d'ethnologie* 3:58–67.
——— 1964. 'La technologie, science humaine', *La pensée* 115:28–35.
——— 1968. 'La technologie culturelle. Essai de méthodologie', in J. Poirier (ed.), *Ethnologie générale.* Paris: Gallimard, pp. 731–822.
——— 1972. 'Souvenirs personnels', *L'Arc* (Marcel Mauss) 48:89–90.
——— 1987. *La technologie, science humaine. Recherches d'histoire et d'ethnologie des techniques.* Paris: Editions de la Maison des Sciences de l'Homme.
Haudricourt, A.-G. and P. Dibie 1987. *Les pieds sur terre.* Paris: Metaillié.
Hecht, J.M. 2003. *The End of the Soul – Scientific Modernity, Atheism, and Anthropology in France.* New York: Columbia University Press.
Herf, J. 1984. *Reactionary Modernism. Technology, Culture, and Politics in Weimar and the Third Reich*, Cambridge: Cambridge University Press.
Hinsley, C. 1981. *Savages and Scientists: the Smithsonian Institution and the Development of American Anthropology 1846–1910.* Washington: Smithsonian Institution Press.

Hobsbawm, E. and T. Ranger (eds) 1983. *The Invention of Tradition*. Cambridge: Cambridge University Press.
Hubert, H. 1903. 'Technologie. Introduction', *Année sociologique* 6:567–68. **Text 2.**
______ 1903. 'Frémont Ch. (1902) Evolution de la fonderie de cuivre (C.R.)', *Année sociologique*, 7:681–82.
James, W. and N.J. Allen (eds) 1998. *Marcel Mauss: A Centenary Tribute*. Oxford: Berghahn Books.
Johnson, D. and M. Johnson 1987. *The Age of Illusion: Art and Politics in France 1918–1940*. London: Thames & Hudson.
Karady, V. 1968. 'Présentation de l'édition', in Marcel Mauss, *Œuvres, vol I*. Paris: Editions de Minuit, pp. i-lviii.
Karsenti, B. 1994. *Le Fait social total*. Paris: Presses Universitaires de France.
______ 1997. *L'Homme total. Sociologie, anthropologie et philosophie chez Marcel Mauss*. Paris: Presses Universitaires de France.
Kuper, A. 1973. *Anthropologists and Anthropology: the British School, 1922–1972*. London: Allen Lane.
Kurasawa, F. 2003. 'Primitiveness and the Flight From Modernity: Sociology and the Avant-garde in Inter-war France', *Economy and Society*, 32/1:7–28.
Labriola, A. 1897. *Essais sur la conception matérialiste de l'histoire*. Paris: V. Giard & E. Brière.
Lacroix, B. 1981. *Durkheim et le politique*. Montréal: Presses de l'Université de Montréal.
Lebovics, H. 1992. *True France. The Wars Over Cultural Identities, 1900–1945*. Ithaca: Cornell University Press.
Lemonnier, P. 1980. *Les Salines de l'Ouest : Logique technique, logique sociale*. Paris: Editions de la Maison des Sciences de l'Homme.
______ 1992. *Elements for an Anthropology of Technology*. Ann Arbor: The University of Michigan.
Leroi-Gourhan, A. 1943. *Evolution et techniques. Vol. 1. L'Homme et la matière*. Paris: Albin Michel.
______ 1945. *Evolution et techniques. Vol. 2. Milieu et techniques*. Paris: Albin Michel.
______ 1964. *Le Geste et la parole*. Paris: Albin Michel.
______ 1982. *Les Racines du monde. Entretiens avec C.-H. Rocquet*. Paris: Belfond, 1982.
______ 2004. *Pages oubliées sur le Japon*. (recueil posthume établi par Jean-François Lesbre). Grenoble: Jérôme Millon.
Lévi-Strauss, C. 1950. 'Introduction à l'œuvre de Marcel Mauss', in M. Mauss, *Sociologie et anthropologie*. Paris: Presses Universitaires de France, pp. Iix-lii.
Lindenberg, D. 1990. *Les Années souterraines, 1937–1947*. Paris: La Découverte.
Lukes, S. 1973. *Emile Durkheim. His Life and Work*. London: Penguin (1988 edition).
Marcel, J.-C. 2001. *Le Durkheimisme dans l'entre-deux-guerres*. Paris: Presses Universitaires de France.
Marin, L. 1925. *Questionnaire d'ethnographie (table d'analyse en ethnographie), Bulletin de la Société d'Ethnographie de Paris*. Alençon: Imprimerie Laverdure.
Mauss, M. 1903. 'Esquisse d'une théorie générale de la magie' (with H. Hubert), *Année sociologique* 7:1–146.
______ 1906. 'Essai sur les variations saisonnières des sociétés eskimos. Étude de morphologie sociale' (en collaboration avec H. Beuchat), *Année Sociologique* 9:39–132.

——— 1913. 'La théorie des couches et des aires de civilisation selon Graebner', *Année sociologique* 12:858.
——— 1920. 'La Nation et l'internationalisme', in *The Problem of Nationality*, Proceedings of the Aristotelian Society, 20:242–51.
——— 1924a. 'Rapports réels et pratique de la psychologie et de la sociologie', *Journal de psychologie* 21:892–922.
——— 1924b. 'Appréciation sociologique du bolchevisme', *Revue de métaphysique et de morale* 31:103–32.
——— 1925a. 'Essai sur le don. Forme et raison de l'échange dans les sociétés archaïques', *Année sociologique* (n.s) 1:30–186.
——— 1925b. 'Frobenius L. – Atlas Africanus', *Année sociologique* (n.s) 1:302–06.
——— 1925c. 'Graebner F. – Ethnologie', *Année sociologique* (n.s) 1:310–18.
——— 1925d. 'Elliot Smith G. – Elephants and Ethnologists; Perry W. J. – The Children of the Sun; Perry W. J. – The Growth of Civilization', *Année sociologique* (n.s) 1:330–42.
——— 1927. 'Divisions et proportion des divisions de la sociologie', *Année sociologique* (n.s) 2:98–176. **Text 5.**
——— 1929. 'Débat sur l'origine de la technologie humaine', *L'Anthropologie* 39:129–30. **Text 6.**
——— 1930. 'Les civilisations. Eléments et formes', in Fondation pour la Science – Centre international de synthèse (eds), *Civilisation. Le mot et l'idée*, 1re Semaine internationale de synthèse, fasc. 2. Paris: La Renaissance du Livre, pp. 81–108. **Text 7.**
——— 1930/1998. 'An Intellectual Self-portrait', in James and Allen (eds), pp. 29–42.
——— 1933a. 'La sociologie en France depuis 1914', *La science française*. Paris: Nouvelle édition Larousse, pp. 36–48.
——— 1933b. 'Intervention à la suite de la communication de J. Piaget: "L'individualité"'. Centre international de synthèse. Troisième semaine internationale de synthèse (1931). Paris: Alcan pp. 118–20.
——— 1934. 'Fragment d'un plan de sociologie générale descriptive. Classification et méthode d'observation des phénomènes généraux de la vie sociale dans les sociétés de type archaïque (phénomènes généraux spécifiques de la vie intérieure de la société)', *Annales sociologiques* série A, 1:1–56. **Text 8.**
——— 1935. 'Les techniques du corps', *Journal de psychologie* 32:271–93. **Text 9.**
——— 1938. 'Une catégorie de l'esprit humain: la notion de personne celle de "moi"' (Huxley Memorial Lecture, 1938), *Journal of the Royal Anthropological Institute* 68:263–81.
——— 1939. 'Conceptions qui ont précédé la notion de matière', in Fondation pour la Science – Centre international de synthèse (eds), *Qu'est-ce que la matière? Histoire du concept et conception actuelle*, XIe Semaine internationale de synthèse. Paris: Presses Universitaires de France, 1945, pp. 15–24. **Text 11.**
——— 1948. 'Les techniques et la technologie' *Journal de psychologie* 41:71–78. **Text 12.**
——— 1947. *Manuel d'ethnographie*. Paris: Payot. **Text 10.**
——— 1950. *Sociologie et anthropologie*. Paris: Presses Universitaires de France.
——— 1953. 'La nation', *Année sociologique* (3rd series), pp. 7–68. **Text 4.**
——— 1968–1969. *Oeuvres*, Vols. I–III. Paris: Editions de Minuit.

______ 1997. *Ecrits politiques*, textes présentés par Marcel Fournier. Paris: Fayard.
______ 2004. *On Prayer* (Introduced by W.S.F. Pickering), Oxford, Durkheim Press, Berghahn Books.
______ 2005. *The Nature of Sociology* (Introduced by M. Gane), Oxford, Durkheim Press, Berghahn Books.
______ 2006. *The Manual of Ethnography* (Introduced by N. Allen), Oxford, Durkheim Press, Berghahn Books.
Meyerson, I. 1946. *La fonction psychologique et les œuvres.* Paris: Vrin.
Mitcham, C. 1994. *Thinking through Technology: The Path between Engineering and Philosophy*. Chicago: University of Chicago Press.
Monnerot, J. 1946. *Les faits sociaux ne sont pas des choses*. Paris: Gallimard.
Montandon, G. 1934. *L'ologénèse culturelle. Traité d'ethnographie cyclo-culturelle et d'ergologie systématique*. Paris: Payot.
______ 1937. *La civilisation Aïnou*. Paris: Payot.
Mucchielli, L. 1998. *La Découverte du social. Naissance de la sociologie en France*. Paris: La Découverte.
Mumford, L. 1934. *Technics and Civilization*. New York: Harcourt, Brace and Co.
Parkin, R. 1996. *The Dark Side of Humanity: the Work of Robert Hertz and its Legacy*. Amsterdam: Harwood Academic Publishers.
Paul, H.W. 1985. *From Knowledge to Power: the Rise of the Science Empire in France*. Cambridge: Cambridge University Press.
Pearce, F. 1989. *The Radical Durkheim*. London: Routledge.
______ 2003. 'Introduction: the Collège de sociologie and French Social Thought', *Economy and Society*, 32/1:1–6.
Pickering, W.S.F. 1984. *Durkheim's Sociology of Religion: Themes and Theories*. London: Routledge.
______ 1994. *Durkheim on Religion*. Oxford: Oxford University Press.
______ (ed.) 2003. *Durkheim Today*, Oxford: Durkheim Press, Berghahn Books.
Prochasson, C. 1991. *Les années électriques (1880–1910)*. Paris: La Découverte.
______ 1993. *Les intellectuels, le socialisme et la guerre, 1900–1938*. Paris: Seuil.
Prochasson, C. and A. Rasmussen 1996. *Au nom de la patrie. Les intellectuels et la première guerre mondiale (1910–1919)*. Paris: La Découverte.
Rivet, P. 1943. *Les origines de l'homme américain*. Montréal.
Schlanger, N. 1991. 'Le fait technique total - La raison pratique et les raisons de la pratique dans l'œuvre de Marcel Mauss', *Terrain*, 16:114–30.
______ 1998. 'The Study of Techniques as an Ideological Challenge: Technology, Nation and Humanity in the Work of Marcel Mauss', in James and Allen (eds), pp.192–212.
______ 1999. 'De la Rédemption à la sauvegarde: contenu et contexte de la technologie du *Bureau of American Ethnology*', in J.-L. Jamard, A. Montigny and F.-R. Picon (eds), *Dans le sillage des techniques. Hommage à Robert Cresswell*. Paris: l'Harmattan, pp. 483–512.
______ 2004, '"Suivre les gestes, éclat par éclat": la chaîne opératoire de Leroi-Gourhan', in Audouze and Schlanger (eds), pp. 127–47.
______ 2005. 'The chaîne opératoire', in C. Renfrew and P. Bahn (eds), *Archaeology: The Key Concepts*. London: Routledge, pp. 25–31.
Schneider, W.H. 1990. *Quality and Quantity: The Quest for Biological Regeneration in Twentieth-century France*. Cambridge: Cambridge University Press.

Sebestik, J. (ed.) 1984. *De la technique à la technologie*, Cahiers STS, n. 2, Paris.
Sigaut, F. 1994. 'Technology', in T. Ingold (ed.), *Companion Encyclopedia to Anthropology*. London: Routledge, pp. 420–59.
Smith, W.D. 1991. *Politics and the Sciences of Culture in Germany, 1840–1920*. Oxford: Oxford University Press.
Sorel, G. 1895. 'Les théories de M. Durkheim', *Le devenir social*, 1:1–7, 22–26.
Spengler, O. 1923. *The Decline of the Occident*. London: Allen & Unwin.
Spengler, O. 1931. *Der Mensch und die Technik. Beitrag zu einer Philosophie des Lebens*, Munich: Beck.
Stanley, J. 1981. *The Sociology of Virtue: the Political and Social Theories of George Sorel*. Berkeley: University of California Press.
Stocking, G.W. 1987. *Victorian Anthropology*. New York: Free Press.
Tarot, C. 2003. *Sociologie et anthropologie de Marcel Mauss*. Paris: La Découverte.
Tombs, R. (ed.) 1991. *Nationhood and Nationalism in France - From Boulangism to the Great War, 1889–1918*. London: HarperCollins.
Trigger, B. 1989. *A History of Archaeological Thought*. Cambridge: Cambridge University Press.
Vatin, F. 2004. 'Mauss et la technologie', *Revue du MAUSS*, 23:418–33 (followed on pp. 434–50 by an annotated presentation of Mauss 1941).
Vogt, P. 1976. 'The Uses of Studying Primitives: a Note on the Durkheimians, 1890–1940', *History and Theory* 15/1:33–44.
Warnier, J.P. 1999. *Construire la culture matérielle. L'homme qui pensait avec ses doigts*. Paris: Presses Universitaires de France.
Weisz, G. 1983. 'The Republican Ideology and the Social Sciences; the Durkheimians and the History of Social Economy at the Sorbonne', in Besnard (ed.), pp. 90–119.
Winner, L. 1977. *Autonomous Technology: Technics-out-of-control as a Theme in Political Thought*. Cambridge, MA: M.I.T. Press.
Winter, J. 1995. *Sites of Memory, Sites of Mourning. The Great War in European Cultural History*. Cambridge: Cambridge University Press.
Zeldin, T. 1977. *France, 1848–1945*. Oxford: Oxford University Press.
Zimmerman, A. 2001. *Anthropology and Antihumanism in Imperial Germany*. Chicago: Chicago University Press.

Index

12. 'Natives "watching Capt. Cook, 1770"'. P.112.ACH1. New South Wales, Australia. Mounted Haddon Collection (CUMAA©). 'In the history of the beginnings of humanity no examples are known of humans who are wholly devoid of tools' (Mauss, text 10).